A Fractal Topology of Time

Deepening into Timelessness

Kerri Welch

Copyright © 2010, 2020 Kerri Welch
All rights reserved.

All rights reserved. No part of this publication may be reproduced, distributed, or transmitted without the prior written permission of the publisher, except brief quotations in noncommercial uses permitted by copyright law.

ISBN: 978-1-7345762-0-7 (paperback)
ISBN: 978-1-7345762-1-4 (ebook)

Second Edition 2020

Published by
Fox Finding Press

TextureOfTime.wordpress.com

Contents

Contents iii
Preface 1
1 Mind and Matter: Quantum Consciousness 5
    1.1   Consciousness and Protein Conformation 7
    1.2   Insight and Non-Computability 9
    1.3   Quantum Mechanics 10
           1.3.1   *Wave-Particle Duality* *11*
           1.3.2   *Superposition* *14*
           1.3.3   *Uncertainty* *15*
           1.3.4   *Non-locality and Quantum Entanglement* *16*
    1.4   Orchestrated Objective Reduction 17
           1.4.1   *Objective Reduction* *17*
           1.4.2   *Orchestrated OR* *18*
           1.4.3   *Subjectivity* *20*
2 Time and Timelessness 25
    2.1   The Limits of Time's Arrow 26
           2.1.1   *Philosophical Limitations* *29*
    2.2   Temporal Symmetry 30
           2.2.1   *Forwards and Backwards, Simultaneously* *31*
           2.2.2   *Quantum Temporal Oscillations* *32*
           2.2.3   *Spacetime Flexibility* *34*
           2.2.4   *Closed Timelike Loops* *37*
           2.2.5   *Counterfactual* *39*
    2.3   The Expanded Moment 43
           2.3.1   *The Speed of Time* *43*
           2.3.2   *The Frozen Timeline of the Block Universe* *45*
           2.3.3   *Temporal Symmetry as a Subset of Timelessness* *47*
           2.3.4   *Time and Timelessness* *48*
           2.3.5   *Subtle Causality* *50*
3 Fractal Time 53
    3.1   Fractals 54
           3.1.1   *Mandelbrot Set* *56*

	3.1.2	Fractal Dimension 59
	3.1.3	Fractal Models of Time 61
3.2		Nottale's Fractal Spacetime 62
	3.2.1	Scale Relativity 64
	3.2.2	Non-differentiability 64
	3.2.3	Fractals in Nature, Optimization 66
3.3		Timelessness and Fractal Dimensionality 68
	3.3.1	Scale Divergence and Uncertainty 68
	3.3.2	Thickening of the Quantum Temporal Dimension 70
	3.3.3	Non-local Coherence 71
	3.3.4	Temporal Reversibility and Timelessness 73
	3.3.5	Fractal Dimension of Time 74
3.4		Imaginary Time 75
	3.4.1	Quantum Mechanics 77
3.5		Vrobel's Fractal Time 81
	3.5.1	Fractal Measures 81
	3.5.2	Timelessness, Insight, and Fractal Prime 83
	3.5.3	Subjective and Objective Time 86
	3.5.4	Temporal Density 88

4 Deep Transcendence 94
4.1 Variables of Subjective Time 94
4.2 Repetition 97
4.2.1 Deepening Time 100
4.2.2 Deepening Cosmological Time 101
4.2.3 Abstraction and Quantum Collapse 104
4.2.4 The Influence of Quantum Timelessness 107
4.2.5 The Spectrum of Consciousness 108
4.3 Attention 111
4.3.1 Frequency 112
4.3.2 Temperature and the Internal Clock 115
4.3.3 EEG and Scale Synchronization 116
4.3.4 Dopamine and the Internal Clock 119
4.4 Intention 123
4.4.1 Desired Time 124
4.4.2 Dopamine and Motivation 125
4.4.3 Grasping/Uncertainty 129
4.4.4 Co-Creating Time 132

References 135
Index 140

A FRACTAL TOPOLOGY OF TIME

Preface

There is a redwood canyon--the floor a rust colored carpet of fallen needles. Above the humus, fat-leafed tangy sorrel congregate, sword ferns hover, and ancestor trees tower, trunks licked by fire. All these denizens drink from the falling water, carving its way to meet crashing waves. The trail into the canyon lilts and curves around sleeping boulders, along the steep sloped sides, crisscrossing the creek, seeking the water's source.

Pausing on a creek-side boulder reveals the faces in the cliff, the people of the rock, nestled in eternal bliss. The churning waters thrum with the rhythmic drumming. The Esselen people loved this land for thousands of years. Their name derives from a phrase meaning "I come from the rock." While historians debate various geographical locations of "the rock," with plenty of candidates along the Big Sur coastline, I suspect a deeper relationship with stone. Their thousands of years deep relationship with this place, cultivated a rich field into which I have stepped, their love thick in the trees and stones that knew them.

In this place--a classroom thousands of years deep--the life force that would unfurl into this book, took root in me. What I knew in words became vision and visceral experience. Energy surged through me. Without moving, I flew toward an ever-receding edge, a crack, a boundary that could never be reached. The closer I got, the more intricate patterns unfurled from the interface, flowing into me, through me, recognizing me.

The universe snuggled up against the place I held open within it. We pressed toward each other, holding each other up. The boundary defined me. Like when you've been reading on the couch for a long time and the cushions have so perfectly conformed to your body that you *don't want to move for fear of disrupting the perfect fit*. Yet from the stillness--flow—the undulations of swimming butterfly, the arch and release of outstretched owl wings, whale tail, and fox brow. The shape of these boundaries felt deeply and essentially me. I flew toward the boundary and it blew me back with its rich unfurling.

Energy flew through the holding pattern that constituted me. I held the shape of the unbridgeable gap from which everything flowed forth. The only word I had to describe it was fractal.

Fractals describe infinite depths built by repeating patterns across scales. They birth exquisite visual textures. The satisfaction of viewing fractal images relates to the satisfaction of being immersed in the natural world because nature is fractal. Nature fills every level of scale with complexity. Why would time differ?

Entering the canyon, I had asked for help with my dissertation topic. I emerged both solidified and baffled. Only many years later, did I stumble upon my original request in a journal and recognize that this undeniably fractal experience had been my answer. Fractals claimed me, luring me to continue to strive to articulate their role in time. The pages that follow record my attempt to wrestle this visionary intuition into scientific and philosophical understanding.

Physics spatializes time, representing it as symmetrical, static, and quantified in predictable intervals. In contrast, humans experience time as flowing, in one direction, and passing at different rates. The contrast of the physical and subjective perspectives suggests that a richer description of time is required to encompass them both.

Living in an age of temporal acceleration increases the importance of how we relate to time. Our health and well-being suffer when abandoned to the accelerating flow. Carving out still spaces within the increasing force of the flow connect us to larger, wiser patterns, like those of stone and trees. Attending to the infinite depths of the present expand the freedom and creativity possible there. This book hopes to expand our vocabulary for talking about ways to enrich, savor, and deepen our experience of time by providing an aesthetically appealing scientific correlate. This book will change how you think about time, and thus how you participate with it.

This book presents a revised edition of my dissertation. I offer it here for those curious about how we might conceive of a fractal model of time. If you prefer a more storied approach and are curious about how these ideas have continued to develop, keep your eye out for my next book, a narrative nonfiction approach with the working title, *The Texture of Time: The Science and Magic of Temporal Perception*. As that narrative version diverged from the original dissertation, Shelli Joye encouraged me to publish the original dissertation in paperback form. So, for those who might appreciate the more bare-bones details, here it is. I included some tidbits not included in the

Preface

original dissertation and adjusted some nuances here and there. The methodology, historical background, and literature review have fallen by the wayside for this version, to get to the meat of the theory in shorter order.

As it stands now, chapter one outlines a model of quantum processes in the brain in their subjective correlates as developed by physicist Roger Penrose and anesthesiologist Stuart Hameroff. Chapter two looks at temporal flow and reversibility, timelessness, and the speed of time through the lenses of entropy, quantum mechanics, and relativity.

Chapter three outlines two approaches to fractal time: astrophysicist Laurent Nottale's fractal spacetime and theorist Susie Vrobel's subjective fractal time. Nottale's scale relativity and fractal spacetime describes the temporal reversibility of fractal spacetime at the quantum level, from which unidirectional time arises in the classical realm. Vrobel aligns fractal measures with human temporal experience, drawing on Penrose's concept of insight and philosopher J.M.E. McTaggart's time series.

Chapter four builds on these theories and considers the variables that effect our experienced rate of temporal flow. Interweaving these theories builds a model of time that deepens into timelessness through the dimension of scale. Along the way we consider the roles of frequency, brain wave states, temperature, and dopamine. We arrive at a deeper understanding of ourselves and our role as co-creators of our temporal landscape.

I am forever indebted to the Philosophy, Cosmology, and Consciousness graduate program at California Institute of Integral Studies, whose magical communion of souls continues to enrich and expand my growth through their openness to many ways of knowing, depth of knowledge, intensity of ethical imperative, and far reaching vision. I would like to thank my dissertation committee: Brian Swimme, Sean Kelly, and Don Salisbury; the Yoga Science Foundation for a small grant toward editing; and all those who have offered feedback and encouraged my progress through the years. Any errors or overly fanciful ideas are my own and beyond the responsibility of any who so graciously assisted me in this endeavor.

Many thanks to my ancestors whose love and material support have allowed me to pursue my unusual curiosities well beyond my heart's content; especially enormous gratitude to my mom, Raylene Welch, marvelous facilitator of independence that she is, to whom I dedicate this work.

1

Mind and Matter: Quantum Consciousness

"Yes!" I did an internal fist pump when I spotted "Metaphysics" as a course offering for the upcoming semester. I know, not everyone gets excited to study space and time, mind and matter, causality, and identity, but this was my jam. When the course catalog arrived in my hands as a high school senior a few years earlier, I had poured over the offerings, circling interesting classes to see which majors they added up to. I triple-circled this class.

Surely metaphysics would not let me down. I had had high hopes for Philosophy of Language too. Yet somehow that class managed to avoid all neuroscience and psychology. How could philosophy, the very foundation of all science, not include any scientific knowledge acquired over the past century? It was like a bad breakup. Since science split from philosophy, philosophy would not design to acknowledge science's existence[1]. Surely this would not be the case for metaphysics.

I had picked up a physics major in addition to philosophy and religion, but physics classes did not make time for pondering the meanings of their findings. I needed a place to mull over the implications of relativity, quantum mechanics, chaos theory, and cosmology. I had high hopes for metaphysics.

---

[1] Environmental Ethics was a notable exception here, drawing extensively on recent scientific research.

Sometimes it feels like high hopes exist simply to be dashed, however. One look at the syllabus told me western philosophy's bitterness at the breakup with science was more deeply entrenched than I had bargained for. Physics ignored human experience and philosophy ignored physics. And they both seemed to deprioritize subjective experience in favor of disembodied logic.[1]

In fact, at times, it seemed philosophy strove to act like science, in an attempt to woo back the powerful young upstart, forgetting she herself was bigger than science. Science did not seem too impressed. But it needed philosophy too—the old, big, holistic philosophy that pondered how to wield power with wisdom and what to do when encountering paradox.

Few bridges appeared between the vacuous mechanics of physics and human subjective experience. Luckily, I had already taken one step onto a wobbly, new rope bridge spanning the chasm. It seemed like it might hold.

My hand hovered over the Jan term registration card. I needed Cognitive Science to continue on track for that minor. But there was another class tugging at me. I couldn't place my finger on why, but whatever "Creating Consciousness"[2] was, I wanted it bad enough to step off the beaten path and find out. Abandoning the potential for a cognitive science minor, I stepped onto the rickety bridge.

We kept dream journals, studied creativity, flow states, and altered states. We read Jung, Csikszentmihalyi, and Grof. We journeyed into ourselves as we studied others' explorations of the psyche. The study of consciousness not only had room for science *and* subjective experience, it was *alive*.

A search for "consciousness grad school" set my sights on California. I found a program[3] that aimed not only to study consciousness, but to consciously participate in its transformation. Acknowledging the ecological tipping point that defines our era, the program committed to cultivate more life-affirming ways of living on earth, drawing on wisdom from many traditions: Eastern, indigenous, feminist, ecological, spiritual, and scientific. I needed this level of holistic engagement.

So why learn physics to transform consciousness? Let's just say there is value in recognizing oneself as a miracle of stardust and woven light, as opposed being defined by what one buys. Of course, there are many ways

---

[1] With the notable exception of phenomenology. ☺
[2] Taught by Jerry Middents.
[3] Philosophy, Cosmology and Consciousness at California Institute of Integral Studies, San Francisco, CA

to get to an expanded understanding of oneself, but it's nice to have them cohere with our physical descriptions of reality.

In addition, shifts in physics often track with shifts in consciousness. As our understanding of the world changes, we change how we engage it. The physics revolutions of the 20th century–relativity, quantum mechanics, and chaos theory--have yet to be integrated into how we understand the world and interact with it. Classical mechanics, the physics of the 18th and 19th centuries, brought incredible insights and prowess in the realm of linear causality. 20th century physics, however, suggests that linear causality is not the only way time works. If this is the case, I might behoove us to understand what we have to work with.

In addition to my future grad school, my online inquiries also led me to an online course in Quantum Consciousness[1]--just the bridge I needed. I enrolled. It did not disappoint.

A model of quantum consciousness proposed by physicist Sir Roger Penrose and anesthesiologist Stuart Hameroff offers one way of connecting subjective and objective time, via the realms of the very small. Here time gets interesting.

Penrose recognizes the unique ability of consciousness to access timelessness via the subjective process of insight utilizing non-computable physical process. Penrose argues that the human ability to access insights unattainable by computation hints at consciousness' utilization of a non-computable physical mechanism to access Plato's timeless realm of ideas.[2]

The question then becomes, "What non-computable process might consciousness utilize to access timelessness?" Penrose explores the non-computability of wave function collapse, closed time-like loops, and fractals.

Through his work with anesthesiologist Stuart Hameroff they identified tubulin, a building block of neurons that quits changing shape under the influence of anesthesia, as a possible location for quantum processes in the brain.

1.1 Consciousness and Protein Conformation

Most people know that neurons are the cells that make up the brain and that the cytoskeleton gives cells their structure. Fewer people, however,

---
[1] University of Arizona, Center for Consciousness Studies, Fall 1999
[2] (Penrose 1989)

know about the composition of the cytoskeleton: a network of microtubules each 25 nanometers in diameter. Microtubules convey physical messages to the nucleus through their vibratory, shimmering activity, which appears to transmit "kinks" of information along their lengths.[1] In seeking a physical location where quantum mechanics might influence how consciousness arises in the brain, Hameroff points out that this microtubule activity ceases under the influence of anesthesia, suggesting its connection to consciousness.

Probing even deeper reveals the inner mechanisms of the microtubule activity. Thirteen columns of proteins, called tubulin layer, brick-like, to build the microtubule's walls.[2] Proteins' change shape, powering the movements of our muscles and the more subtle movements of mind as well. Chains of amino acids fold in on themselves to create three-dimensional tubulin that changes shape over a billion times a second. In the center of this structure, regulating the shape changes, lies a hydrophobic pocket. Hydrophobic means "fear of water;" these pockets have a solvency similar to olive oil.

The most potent general anesthetics, dissolve in oil not water/blood, so can penetrate the hydrophobic pocket and impair the ability of the tubulin to change shape. Anesthetics ability to "turn off" consciousness without impairing other brain functions, implicates the hydrophobic packet and the tubulin shape shifting in the functioning of consciousness.

Water and oil don't mix because oil's non-polarity does not interact with water's polarity. Like oil, the molecules within amino acids are generally non-polar, and thus hydrophobic. Even within a non-polar substance, however, subtle shifts in polarity arise as negatively charged electron cloud's density shifts around the positively charged protons within atoms' nuclei. These subtle, rapid shifts in polarity, called dipole couplings or van der Waals forces, interact with the polarities of surrounding molecules forcing the entire protein to restructure and change shape. Thus, the electron movement within the amino acid groups that make up the hydrophobic packet regulate shape changes. The weakest and most numerous of these dipole interactions are the London dispersion forces, which regulate protein folding and conformation.

Anesthetics disrupt London dispersion forces preventing tubulin from changing shape and microtubules from sending messages, thus linking

---

[1] (Hameroff 2009a)
[2] (Hameroff 1999)

these processes to consciousness. When an anesthetic molecule enters a tubulin's hydrophobic pocket, it interferes with the ability of the tubulin's electrons to move around, perhaps through its own London forces. With electron mobility lowered by the influence of anesthesia, the tubulin quit changing shape and the microtubules stop vibrating, shimmering, and transmitting their kinky messages.

The potency of psychedelics also depends on their oil solubility. However, their potency depends on their ability to increase, rather than decrease, electron mobility. Increased electron mobility increases tubulin activity and the possibility of superposition, as discussed in future sections.

Anesthesia inhibits consciousness and tubulin shape shifting, and psychedelics open alternative modes of conscious perception and increase tubulin activity.[1] Thus, tubulin seem strongly implicated as a physical correlate of consciousness, where mind and matter meet.

1.2 Insight and Non-Computability

Might tubulin transformation facilitate consciousness by playing the role of a bit in a computer system? Tubulin shift between two possible shapes, similar to the way computing bits shift between ones and zeros. Computing power, however, has inherent limitations when compared to the brain. Penrose uses the Turing halting problem and Gödel's theorem to illustrate that the unique human capacities of insight and meaning allow us to know things that are not provable, thus establishing computability as a subset of truth, rather than a defining feature. Gödel shows that all systems of formal logic can have propositions, which are both true *and unprovable*. Thus, knowledge based solely on the processes of logic, is incomplete.[2] The key point here is that the human mind can ascertain the truth-value of certain propositions which computability cannot. Thus, Penrose's asserts that consciousness relies on non-computational processes.

Insight, for Penrose, differentiates consciousness from computability. Many people arrive at mathematical truths through seemingly spontaneous strokes of insight, like Archimedes, "Eureka!" moment. Penrose sees mathematical, musical, and creative truths as existing in Plato's timeless realm of ideas. Thus, for the human mind taps into these concepts by tapping into timelessness. Penrose identifies insight as an ability to draw

---

[1] (Hameroff 1999, Sec. 4; 2006)
[2] (Penrose 1989, 105-108)

truths from a realm of timelessness into implementation in the temporal flow of our daily lives. Penrose then asks:

> But even if we accept that consciousness itself has such a curious relation to time–and that it represents, in some, sense, contact between the external physical world and something timeless–how can this fit in with a physically determined and time-ordered action of the material brain?[1]

Penrose explores various non-computational processes that might facilitate consciousness's contact with timelessness. Penrose and Hameroff focus on quantum processes in the brain, but Penrose also highlights the non-computability of the Mandelbrot set which offers a tie in for fractals as we will explore in later chapters. Penrose touches on the power of fractals' non-computability in his treatment of the Mandelbrot set, but does not develop their relationship to time, as this book will.

Penrose, however, presses onward, past the tantalizing mathematics of fractals, in search of other more physical seats of non-computational processing. He proceeds based on the logic that any non-computational processes of the brain, which might orchestrate consciousness "would also have to be inherent in the action of inanimate matter." Mind distinguishes itself from matter through a unique organization "to take advantage of non-computable action in physical laws, whereas ordinary materials would not be so organized."[2] Thus, to explain consciousness, Penrose passes over neurobiology and chemistry and heads straight for the fundamentals of how spacetime interacts with the brain. To do so, he ventures into the realms of quantum mechanics and general relativity.

If quantum processes do somehow play a role in consciousness's ability to access timelessness, what might that look like? Quantum mechanics dives into the tiniest crevices of space and time to see what they hold.

1.3 Quantum Mechanics

Quantum computing is one of the next frontiers of computer science and, for Penrose's purposes, an excellent candidate for a neurological and cosmological basis for accessing non-computability. Quantum

---

[1] (Penrose 1989, 446)
[2] (Penrose 1994, 216)

computing employs the quantum mechanical principles of superposition, non-local quantum entanglement, and subsequent collapse of a wave function. Superposition extends the traditional binary bit system of ones and zeros to a qubit system in which the superposition of one and zero provides a third option, and possibly some more mysterious advantages via quantum entanglement. The link to non-computability enters through Penrose's theory of quantum gravity's role in collapsing the wave function. Let's begin with some of the foundations of quantum mechanics: wave-particle duality, the Schrödinger equation, the Heisenberg uncertainty principle, and the Einstein Podolsky Rosen (EPR) paradox.

1.3.1   Wave-Particle Duality

Classical mechanics described the macroscopic motion of objects on earth and in the heavens for many years before quantum theory began describing the small-scale activities of matter and energy in the early 20th century. The discovery that energy propagates both in quanta, or in discrete units, and continuously, as a waveform, is the foundation for quantum mechanics and the many mysteries that unfold from there.

Early in the 19th century, the double slit experiment performed by Thomas Young bolstered the case for the wave theory of light. When shining a beam of monochromatic light through two small slits in an opaque barrier, one sees, not a two-slit projection, as one would expect, but an interference pattern of alternating dark and light regions on the screen beyond the barrier.

Interestingly the brightest patch falls directly behind the space where the light was blocked between the two slits. This implies the interference of waves, the amplification and dampening due the crests and toughs of the waves reinforcing or negating each other.

The continuity of energy's waveform, perhaps, seems intuitive. However, the argument for energy's particulate nature gains traction about a century later. Three phenomena of the late 1800s and early 1900s, cathode rays, the photoelectric effect, and blackbody radiation, found eventual explanation through the quantization of energy in the early 1900s. Cathode rays revealed electrons as negatively charged, particulate, carriers of electricity. The photoelectric effect describes the ability of light to release

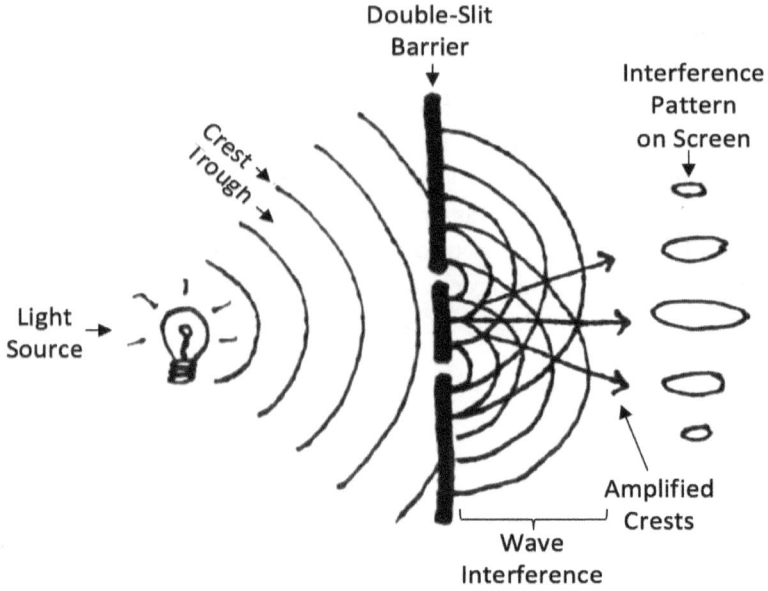

Figure 1.2 Double Slit Experiment

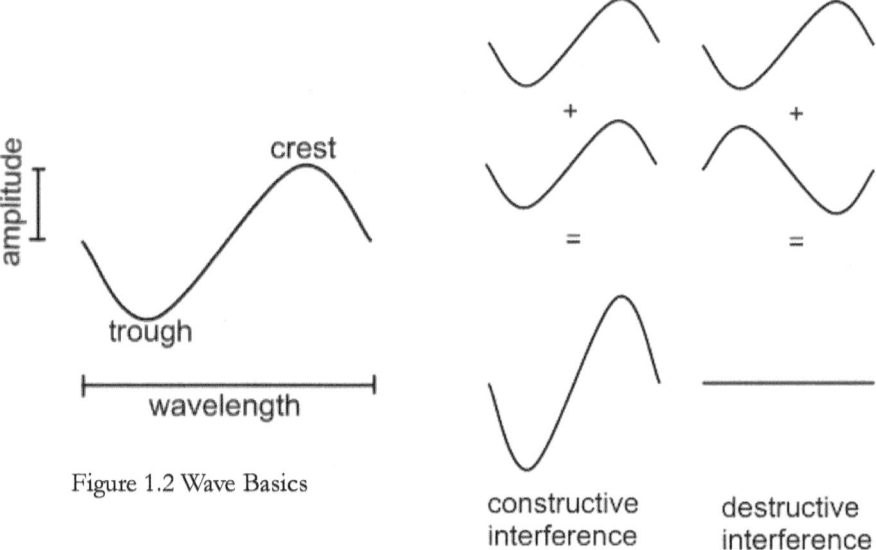

Figure 1.2 Wave Basics

electrons from matter. It led to Einstein-Maric's[1] 1905 theory of quantized light, or photons. Einstein-Maric used Planck's law,

$$\varepsilon = hf$$

to explain the photoelectric effect through the quantization of light. This established Planck's constant, h, as the constant of proportionality between ε, the discrete energy packets, and f, the frequency of radiating energy. Planck's Law finally accurately explained blackbody radiation, avoiding the erroneous classical prediction of ultraviolet catastrophe.[2]

However, if energy is particulate, what does that mean for the wave theory of light as evidenced in the interference patterns of the double slit experiment? Here the plot thickens. Even with a light source so faint that only one particle goes through the barrier at a time, the interference pattern remains. The wave does not arise as an epiphenomenon of a collection of photons, but as a property of each individual photon. Each single photon somehow interferes with itself. Adding additional mystery, if a measurement device tracks which slit the photon goes through, then the interference pattern disappears. The act of measurement changes the behavior of system. No longer can scientists assume that they can be objective observers, for their observations influence that which they observe. We are not separate from the reality we observe.

Louis de Broglie's 1924 doctoral thesis established the notion of wave-particle duality by combining Einstein-Maric's relation of mass to energy, $E=mc^2$, with the energy to frequency relation $\varepsilon = hf$, to show that mass also has wave characteristics. A particle's de Broglie wavelength is Planck's constant divided by the particle's momentum.

From de Broglie's theory of the wave properties of single particles, Erwin Schrödinger developed the Schrödinger equation in 1926, describing the wave nature not only of particles, but also of systems of any size. The next year Werner Heisenberg published the uncertainty principle which states that the more precisely momentum is known, the less precisely position is known and visa-versa. Then in 1935, Einstein, Podolsky, and Rosen developed a thought experiment known as the EPR paradox, that they hoped would demonstrate the incompleteness of quantum theory.[3]

---

[1] I will refer to the work of the 1905 *annus mirabilis* as authored by Einstein-Maric, in acknowledgement of Einstein's intimate collaboration with Mileva Maric, a physicist of equal training and his wife from 1896 – 1916.
[2] (Penrose 2005, 501-503)
[3] (Einstein et al. 1935)

Experimentation, however, upheld the quantum predictions and violated classical logic. The next three sections will go into more detail on the last three developments: the Schrödinger equation, Heisenberg's uncertainty principle, and the EPR paradox.

1.3.2   Superposition

The Schrödinger wave equation solutions are wave functions, Ψ, that each describe a different possible state for the system. For example, "Schrödinger's cat," when unobserved in a closed box, can be in one of two states, awake or asleep, or analogously, a particle can be spin up or spin down. Each state is associated with a wave function. The nature of the Schrödinger equation is such that the sum of any number of these solutions is also a solution. Thus, until it is determined which of the states a system is actually in, the system is in a superposition of possible states, the cat is both awake and asleep; the particle's spin is both up and down. Once measured however, the superposition of wave functions "collapses" into only one of the possible states. Measurement magnifies a quantum state to a classical state.[1] One opens the box to reveal an alert or napping cat, a particle passes through a detector which measures the spin as up or down.

There is no way to determine, with 100% certainty, into which state the system will collapse. One can only determine the probable occurrence of each possible state. For example, if a cat sleeps 18 hours out of every 24-hour day, then there is a 75% chance opening the box will reveal a sleeping cat. Max Born interpreted Schrödinger's equation as the probability amplitude. As it turns out, the square of the absolute value of wave function's amplitude gives the probability of a system collapsing into the state associated with that wave function. Somehow, probability governs the translation from quantum to classical, though many scientists, Einstein and Schrödinger included, found it unacceptable that an underlying determinism had yet to displace the whims of probability in describing reality.

Whether superposition describes reality or the state of our knowledge, remains unclear. The non-deterministic nature of a probabilistic collapse of the wave function puts the quantum collapse in the running for a non-computable mechanism in which to base consciousness. Penrose proposes that the collapse of the wave function might play a part in the non-

---

[1] (Penrose 1994, 263)

computability of consciousness. He looks to the brain's tubulin as a possible seat for quantum processing.

Quantum computing operates on a *qu*bit system, instead of a typical computational bit system. A qubit includes two possible states, similar to a traditional bit system of ones and zeros, but also includes the possibility of *the superposition of those two states*. Penrose, through work with anesthesiologist Stuart Hameroff, suggests that tubulin proteins might operate like quantum computers, shifting between their two shapes and utilizing the limbo of these superposed states to explain the edge that consciousness has over computation. When anesthesia interrupts the London dispersion forces in the brain, a protein can no longer change shape and thus can no longer participate in a superposition of states. These forces also affect the time spent in superposition, which has interesting possible correlations with subjective states of consciousness, as explored in later sections.

1.3.3   Uncertainty

Another of the mysteries of quantum mechanics is Heisenberg's Uncertainty Principle. The more precisely one knows the position of a particle, the less precisely one knows its momentum and vice versa.

Whenever we observe/measure a system, we alter it simply by making the observation. We can never know how a system would act independent of an observation. To get an exact position measurement one must use light with a short wavelength and thus high momentum, which drastically alters the momentum of the particle because of the energy imparted to it. In order to measure the momentum, one must use light of low momentum, and thus long wavelength, which cannot accurately measure position.

The uncertainty equation states that the change in momentum, P, times the change in position, X, must be greater than or equal to the reduced Planck constant, $\hbar$, over 2.

$$\Delta P \Delta X \geq \hbar/2$$

ΔP is the range of possible momenta. The smaller the range, the more accurately we know the momentum. For greatest accuracy $\Delta P = 0$, thereby knowing the particle's exact momentum. The same applies for position, X. To know either exactly would yield a zero product, violating the inequality. Therefore, uncertainty remains unavoidable in a particle's position and momentum. The Schrödinger wave equation's inability to predict particle locations, with more precision than the probability of a particle's occurrence in a specific location, stems from this uncertainty.

Einstein believed that beneath our inability to know, reality still has exact, and simultaneous, position and momentum measurements. Bohr, and the supporters of the Copenhagen interpretation, on the other hand, argued that we can only predict the outcome of measurement, the reality of position and momentum beyond the quantum threshold exceeds our knowledge and may not even exist.[1]

1.3.4   Non-locality and Quantum Entanglement

In 1935, dissatisfied with the limitations on describing reality imposed by the uncertainty principle, Einstein, Podolsky, and Rosen (EPR) put forth an argument to prove the incompleteness of quantum theory, because it cannot completely describe non-commuting variables, such as position and momentum of the uncertainty principle, simultaneously.[2] The experiment that finally vindicated quantum theory involved a particle that decays into two entangled particles with correlated behavior. EPR claimed that hidden variables, which quantum mechanics does not account for, establish the incompleteness of quantum mechanics.

The EPR thought experiment goes like this: a particle in a zero-spin state breaks down into two electrons of opposite spin states (one up, one down) moving away from each other. The experimenters do not know which particle spins up or down until they measure one particle. A Stern-Gerlach device set in one particle's path measures its spin state. Once we know the spin of one particle, then we know the other particle's spin must be the opposite because of their inverse correlation. EPR argued that the measured particle could not possibly have instantaneously communicated the measurement results to the other particle, so the other particle must have had the opposite spin state all along, even if we could not detect it. If spin is a local hidden variable, then incompleteness falls on the theory of quantum mechanics rather than on quantum reality itself. This was the initial EPR argument.

---

[1] (Hebert 1985, 23-24)
[2] (Einstein et al. 1935)

In 1964, John Bell showed that if local hidden variables were in fact responsible for this entanglement, then the experimental results should conform to certain probabilistic rules expressed by Bell's inequality. Once performed, however, the experiment's results violated Bell's inequality, which would have held if local, definite valued hidden variables existed. In other words, the particles do not have a definite spin until measured, so how do they maintain their opposite stance? Herein lays the paradox. How do the particles know instantaneously what the other is doing even when separated by a substantial distance?

David Bohm, however, devised a system based on non-local hidden variables, as opposed to the local hidden variables that the violation of Bell's theorem disproved. He postulated that each particle was merely a different perspective of a larger reality. He referred to this larger reality as the implicate order, an underlying system from which all the things of the material world (the explicate order) manifest momentarily. Bohm theorized a particle field known as quantum potential that does not depend on distance from the particle, and therefore extends infinitely. One particle's quantum potential interferes with other particles' quantum potentials. The changes in the original particle's field convey information about the other particles, acting like radar reporting about its surroundings.[1] Taking perspectives on timelessness seriously, as this book explores, may offer new perspectives on the EPR paradox. Similar to Bohm's perspective, this book posits timelessness as a global hidden variable of the underlying implicate order.

1.4 Orchestrated Objective Reduction

1.4.1 Objective Reduction

In the previous section, I briefly touched on the act of measurement facilitating the collapse of the wave function. Penrose proposes that when left to its own devices a system in superposition will eventually collapse into one state or another, even if no one is measuring it.[2] He calls this the objective reduction (OR) of the wave function and postulates that quantum gravity causes the self-collapse. He suggests that OR is non-computable, being neither algorithmic nor random, but rather, depends on the non-local

---

[1] (Bohm 1990)
[2] (Penrose 2004)

influences of quantum gravity and entanglement. He hopes to describe consciousness's ability to contact timelessness via non-computable process, and by choosing OR as a possible mechanism suggests the timelessness of the quantum realm. Since we have no accepted theory of quantum gravity yet, this is highly speculative. I have chosen to entertain his speculations as they provide an important starting point for trying to imagine what role quantum mechanics might play in consciousness.

Penrose extends general relativity to the quantum level recognizing that every mass has an associated spacetime dent. He suggests that when two superposed states exist simultaneously, their spacetime dents form something of a spacetime blister. The blister expands as the states pull away from one another, until reaching a point where the gravitational attraction between the two states collapses their superposition and snaps the system into one state or the other.

This decoherence threshold derives from the uncertainty principle for time and energy, which mirrors that of position and momentum,

$$\Delta E \geq \hbar/2\Delta t$$

In this formula, E is the energy contained in the gravitational attraction between the two superposed masses; $\hbar$ is Planck's constant, h, divided by 2π; and t is the time it takes for the superposition to collapse. The greater the mass, the greater the quantum gravity pulling the masses together, the less time the object can remain in a state of superposition. The larger the mass, the faster the wave function collapses. The smaller the mass, the longer the object can remain in superposition. So, a one kg mass would last about 10^{-37} sec, whereas a tubulin could maintain superposition for about one second if undisturbed.

1.4.2   *Orchestrated OR*

In 1994 Penrose teamed up with anesthesiologist Stuart Hameroff to link his OR theory to physical processes in the brain. The Penrose-Hameroff model of Orchestrated Objective Reduction, Orch OR, identified tubulin states as the qubits of the brain's quantum processing.[1] The "orchestrated" part comes in when tubulin's quantum states entangle with

---

[1] (Penrose 1994)

one another in growing numbers until the superposition of the combined tubulin collapse.

EEG measures the frequency of electrical activity in the brain ranging from the 0-4 Hz delta waves of deep sleep to 40+ Hz gamma waves, which seems to bind together different aspects of experience when synchronizing across the brain. "Gamma synchrony/40 Hz coherence was found to correspond with attention, face and linguistic recognition, visual binding, task performance, working memory, dreaming and consciousness (e.g., by virtue of its selective disappearance with general anesthesia)."[1] The 40 Hz cycling seems unrelated to axon firings, but does match up with dendrite gap junction firings that link many neurons in a common membrane, allowing them to act as a single neuron. Cytoplasmic cycling between liquid and solid states in sol-gel transitions can occur at 40 Hz and may provide a mechanism for alternately protecting and communicating the results of tubulin superpositions.[2]

This 40 Hz cycling could then correspond to a conscious "event" every 25 ms. Remarkably, this is on the same order as the thirteen and twenty conscious moments, or thoughts, per second as noted in various Buddhist writings.[3] Hameroff also makes the comparison with philosopher Alfred North Whitehead's actual occasions of experience.

According to OR equation, it takes about 2×10^{10} entangled tubulin in superposition to correspond to the gravitational energy necessary to facilitate an Orch OR conscious event every twenty-five ms, corresponding to an excited EEG state of 40 Hz gamma waves. Figure 1.3 illustrates this.

---

[1] (Hameroff 2009b).
[2] (Hameroff 1999; 2009a).
[3] (Hameroff and Penrose 2009).

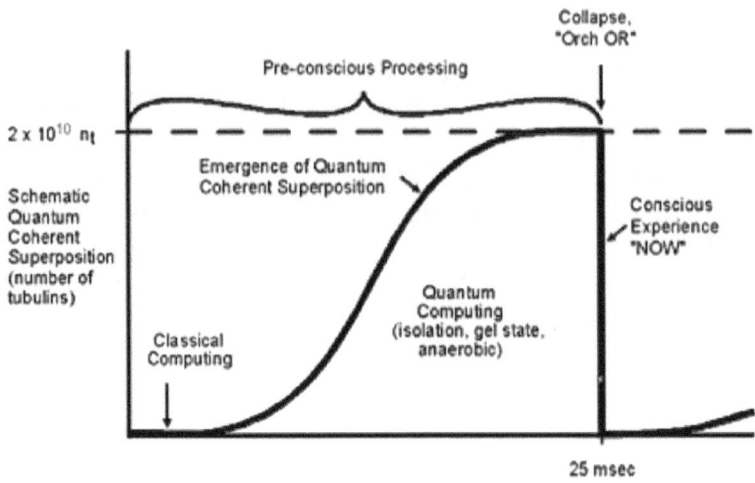

Figure 1.3 Conscious moment as a function of number of tubulin in superposition.[1]

1.4.3 Subjectivity

Hameroff and Penrose's Orch OR shows one way that the non-computability of quantum processes relates to consciousness. This model is important not only for its ability to link physics and biology, but also for its links to subjective experience.

Imagine you are at a restaurant trying to decide what to order, portabella fajitas or fish tacos. Your mind, a realm of infinite possibility, can imagine how either option might play out. In this mental realm, you are free from the constraints of real consequences. One hallmark of this realm is temporal reversibility. You can play out one decision in your mind and then go back and play out the other decision, comparing which one seemed more satisfying. The time spent in this realm of possibility allows you to make a final decision that you will then play out in the external macroscopic world. Alternatively, perhaps there are external constraints that demand you make a decision at a specific time, such as the waiter's arrival. Similarly, the tubulin in superposition are in both possible shapes at once until they collapse by the gravitational weight of their own indecision (OR) or are forced into collapse by external interactions (measurement collapses the wave function).

---

[1] Image from (Hameroff, 2009a)

Mind and Matter: Quantum Consciousness

Hameroff extends this analogy to describe how Orch OR might occur during different states of consciousness, based on the biochemistry and the subjective experiences of these states. Building from Figure 1.3, the next figure illustrates the number of tubulin in superposition on the vertical axis and time on the horizontal axis. These graphs display various mental states as they might correspond to the frequency of quantum collapse.

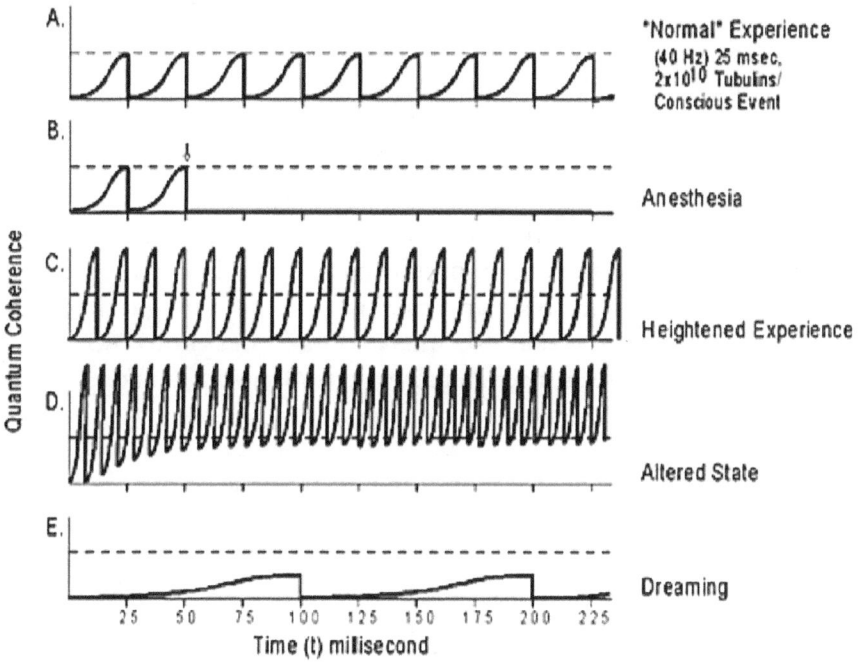

Figure 1.4 States of consciousness as a function of tubulin superposition.[1]

Graph A represents a normal state of consciousness, with an Orch OR conscious event about every 25 ms. This graph simply extends Figure 1.3. Graph B represents the influence of anesthesia as it prevents tubulin conformation and superposition by interfering with the electron mobility of tubulin's hydrophobic pockets. Graph C represents a heightened state of awareness where increased sensory input increases the number of tubulin in superposition, thus achieving Orch OR conscious events at greater frequencies according to Penrose's equation for time spent in superposition varying inversely with mass/energy. These increased frequencies also

---
[1] Image from (Hameroff, 2009a)

correspond to the increased EEG frequencies associated with heightened states.

Graph D represents an altered state as might be associated with meditation or psychedelics. Psychedelic potency has to do with the drug molecule's electron energy available to for transfer to the protein receptor. As anesthesia inhibits electron mobility and thus protein conformation and consciousness, psychedelic drugs increase electron mobility, the rate of protein folding, and the likelihood of superposition.[1] Therefore, this graph suggests that during an altered state of consciousness tubulin enter superposition more quickly, bringing normally subconscious processes into conscious experience. In Graph E, Hameroff speculates that dreaming consists of fewer tubulin in superposition over longer periods of time, corresponding with sleep's decreased EEG frequency of 10 Hz or lower.

It is not a new idea that quantum mechanical processes might play some role in the processes of consciousness. Niels Bohr recognized that quantum processes likely play an essential role in thought because the amounts of energy involved are so small.[2] Bohm goes on to recognize that if brains operate quantum mechanically, then observing thinking processes could provide a unique perspective on quantum processes:

> If it should be true that the thought processes depend critically on quantum mechanical elements in the brain, then we could say that thought processes provide the same kind of direct experience of the effects of quantum theory that muscular forces provide for classical theory... the behavior of our thought processes may perhaps reflect in an indirect way some of the quantum-mechanical aspects of the matter of which we are composed.[3]

In quantum mechanics, once you measure a particle, the properties you sought to measure have changed. The mind responds in a similar manner to a particle, when observed. Observe your thoughts and they are no longer the thoughts you sought to observe.

> If a person tries to observe what (s)he is thinking about at the very moment that (s)he is reflecting on a particular subject, it is generally agreed that (s)he introduces unpredictable and uncontrollable

---

[1] (Hameroff 1999, Sec 4).
[2] (Bohr 1934)
[3] (Bohm 1951, 172)

changes in the way their thoughts proceed thereafter.[1]

What applies to the smallest particles, and to our most intimate thought processes, also seems to apply to our largest concepts of reality–God and life.

> as long as we try to grasp God, we shall never realize God. Life itself, as we experience it moment by moment, proceeding as it does directly from God, is the perfect analogy of this truth, for to grasp life is to kill it, or rather, to miss it, and more than ever is this true of God - the Life of life. Pluck a flower and it dies. Take up water from the stream, and it flows no longer. Pull down the blind, but the sunbeam is not trapped in the room. Snatch the wind in a bag, and you only have stagnant air. This is the root of every trouble: man loves life, but the moment (s)he tries to hold onto it (s)he misses it. The fact that things change, move, and flow is their very liveliness, and the harder man hangs on to his life, the sooner (s)he dies of worry.[2]

Is there a common root to the transience of: quantum reality, consciousness, and life? Could the parallels between the flow of our subjective, felt experience and the flow of energetic of quantum reality indicate something of the nature of their deeper interconnection?

In Watts' last sentence above, he refers to the phenomena of change, movement, and flow. It seems that time is the force behind these phenomena, as well as the root of humanity's discontent. Though he does not include the slipperiness of quantum phenomena specifically, how it seems to fit the same pattern, as evidenced in Heisenberg's uncertainty principle. The attempt to stop flow in order to observe or define it inevitably comes up empty handed.

Hameroff argues that time does not propel the flow of consciousness, but that consciousness creates time.[3] Each conscious event reconfigures quantum spacetime geometry through the non-algorithmic process of quantum collapse. Because the process is non-algorithmic, it is non-deterministic and thus non-reversible, ratcheting time forward and

---

[1] (Bohm 1951, 169)
[2] (Watts 1947, 92)
[3] (Hameroff 2003)

creating the unidirectional flow which we experience.[1] I wonder, however, if consciousness *causes* this, as suggested by Hameroff, or *is caused* by this.

Inevitably, looking for a root "cause" will depend on the "notions" of consciousness and the "causality" of time. The deep intertwining of time and consciousness makes it nonsensical and circular to try to prioritize one over the other.

Their interdependence, however, is unmistakable. "One of the most striking and immediate features of conscious perception is the *passage of time*."[2] Consciousness perceives time by remembering and predicting. Both of these are essential components of self-reflexive consciousness. Consciousness also depends on its relationship to time, as seen through Penrose's prescient attention to consciousness' ability to access timelessness through insight.

This chapter explored one possible physical basis for consciousness' access to timelessness, through the quantum superposition of neural proteins. Continuing to build on Penrose's thought in the following chapter, I explore the physics of time and timelessness via thermodynamics and relativity, in an effort to reconcile them to one another, and to subjectivity.

---

[1] (Hameroff 2003, 87)
[2] (Penrose 1994, 384)

2
Time and Timelessness

Our current physical theories of time leave much to be desired. They lack a link to our experience of time. "what our best theories *do* say is almost in flat contradiction with what our perceptions seem to tell us about time . . . In fact, it is *only* the phenomena of consciousness that requires us to think in terms of a 'flowing' time at all."[1] Classical physics describes time as static, symmetrical, and made up of predictable intervals. In contrast, humans experience time as unidirectional, flowing, and passing at different rates. Additional complexities arise in quantum mechanics and relativity with varying speeds of time, temporal reversibility, and timelessness. The contrast of these perspectives suggests that a richer description of time is required to encompass these realities.

One way toward a richer description of time is to integrate a few of the trickier aspects of time: temporal flow, temporal symmetry, timelessness, and the speed of time. Physics addresses these through three different concepts: entropy, symmetry, and relativity. Entropy implies the unidirectional flow of time. Symmetry implies a reversal of this flow. Relativity implies variations in the rate of time and offers the notion of timelessness as a fundamental aspect of the universe. Each of these ideas seems to describe a different reality. Often people assume that one must be more true or fundamental than another. I suggest that each perspective provides a unique and essential take on reality and the truth lies not in which one is more fundamental, but in delineating the scope of each and determining how they fit together.

I begin with entropy, the arrow of time implicit in the second law of thermodynamics, since it corresponds most closely with our everyday experience of time. Our experience of the unidirectional flow of time

---

[1] (Penrose 1994, 384)

provides a context beyond which it is difficult to see or articulate, but science suggests this is merely one-way time manifests in the universe. Beyond the second law, the rest of physics tells another story by obeying temporal symmetry. I will explore interpretations of several physical scenarios of space time-curvature and particle interaction that suggest possible manifestations of time reversal. Then, I consider the implications of relativity, particularly the speed of time as it varies according to relative velocity of inertial frames, and the notion of a determined, frozen, "block" universe.

2.1 The Limits of Time's Arrow

The second law of thermodynamics states that heat flows from hot to cold regions, describing the tendency for energy to spread out and cool off. Another way of saying this is that "entropy" will always increase. An increase in entropy is often defined as an increase in manifest disorder,[1] randomness,[2] or as a loss of information.[3] Think of this as the tendency of life toward disarray. I can blame the second law of thermodynamics for the tendency of my desk to get messy, my car to need repairs, and my body to fail me with greater frequency as I age.

People often assume this law applies universally because it seems to agree with our perception of uni-directional time flow, i.e. you can't put a broken egg back together. However, no other law of physics cares which direction time moves; their equations work equally well forwards and backwards in time. The claim that disorder and randomness are the ultimate destiny of the universe may seem a bit counter-intuitive, not to mention disheartening, given that most of our human goals aim for the opposite. In light of entropy, the very struggle to keep our bodies fueled and maintained to prevent wasting away, simply prolongs our inevitable demise to maximum entropy. Even the act of cleaning, an attempt at lowering entropy/increasing order, releases so much energy in heat, that the corresponding entropy increase is positive.

The universalization of entropy seems to commit a fundamental fallacy, however. "The basic objection to attempts to deduce the unidirectional nature of time from concepts such as entropy is that they are

---

[1] (Penrose 1989, 308)
[2] (Penrose 2005, 690)
[3] (Bohm 1986, 180)

attempts to reduce a more fundamental concept to a less fundamental one."[1] While the second law describes a temporal texture we can relate to, there are some significant limitations to its ability to describe the ultimate reality of time.

While there are many philosophical limitations implicit in the second law, first I will address its more explicit limitations. Specifically, the second law applies to statistical tendencies of macroscopic, closed systems in terms of equilibrium states. This law does not describe the entirety of reality, but a specific portion of reality. Several limitations intertwine here.

First, statistical laws, such as the second law of thermodynamics, predict only scenarios of high probability. Entropy describes probabilities/tendencies rather than describing individual parts within the whole. While improbable states, such as life and consciousness, do happen, their unpredictability relegates them to the periphery. For example, entropy ignores that all of our scientific observations couch within the improbable, perceiving human. The beauty of science lies in its ability to abstract general laws that apply across a broad swatch of scenarios. The exceptions to these rules, however, are as important as the rules themselves.

Ilya Prigogine, a 20th century, Belgian, physical chemist, tackled this problem with his work on far-from-equilibrium dissipative structures, where instances of order (e.g., life) arise within an overall tendency toward disorder.[2] He found that organized energy dissipates more energy than disorganized energy, so that organization contributes to disorganization. Consider how water, by organizing itself into a whirlpool, moves faster down a drain than it would if not organized. Likewise, the ordered structures of life operate at a controlled burn, dissipating energy faster than it would without an organized structure. We are whirlpools of energy, breaking down the gradient between hot and cool as whirlpools in water break down the gradient between fast and slow. Thus, while disorder is the probabilistic tendency of an entire closed system, it is not the rule for every micro-system within the larger system. Some of those microsystems approach disorder faster through organization. Prigogine's study of dissipative systems brings the disobedient micro-instances of order into alignment with the second law, but also validates the tendency toward order as an equal partner in the tendency toward disorder, rather than its antithesis. Order and disorder require each other.

---

[1] (Whitrow 1980, 338)
[2] (Prigogine and Stengers 1984)

David Bohm also offers more nuance to entropy's definition, "A state of high entropy is one in which large micro-differences correspond to little or no macro-differences or, in other words to a state in which micro-information is 'lost' in the macroscopic context."[1] For example, if you stir up a muddy puddle of water, totally rearranging all of the molecules that make it up, it is still a muddy puddle of water. It would be very difficult to tell the difference between the water before and after stirring. There are many varieties of microstates that look like the same macrostate. Large micro-differences correspond to little or no macro-differences in a state of high entropy. These states, where little changes make little difference, are very common.

Less common are highly ordered or complex states in which small micro-differences lead to large macro-differences as described by chaos theory's butterfly effect, where the flap of a butterfly's wing in Texas can causes a typhoon halfway around the world. This seems to suggest that order arises out of chaos, albeit more rarely, in addition to disseminating into chaos. This illustrates how statistical laws, such as the second law of thermodynamics, are limited to describing scenarios of high probability. Without considering the highly improbable as well, the law is incomplete. Just because it is improbable, does not mean it does not occur.

Second, the second law applies to closed systems, another convenient scientific abstraction. In a closed system energy does not enter or leave. Since energy is always moving, imagining it as contained presents a large assumption. The universe as a whole may present such a system; this is our best guess, but still a large assumption. Third, the second law describes states, not the transitions between these states. Since time deals with change, a law that does not describe change, does not describe the reality of time.

Third, the second law describes states, not the change between these states. Since time inherently deals with change, a law that does not describe change does not describe the reality of time adequately. While the second law is helpful and accurate in our everyday lives, it falls short of providing an encompassing description of the reality of time.

Recognizing these limitations, the second law illustrates an exception or special instance, within the larger reality described by the laws of physics. Contrary to the second law of thermodynamics, most physical laws are time symmetrical, functioning both forwards and backwards in time. If the second law is a special instance rather than a universal law, our perception

---

[1] (Bohm 1986, 181)

of temporal flow might also fit as a special instance within an overall temporal symmetry?

2.1.1 Philosophical Limitations

There are several, more philosophical, limitations implicit within the second law. First, while the second law may describe one aspect of our experience of time, it does not describe the entirety of our experience. Humans experience variations in time when it seems to "drag" or "fly." Since we humans experience time passing at different rates depending on our mental state, it seems plausible that the apparent flow of time could be a function of our perception rather than an external reality independent of perception. Since we experience our lives as unfolding linearly in time, we tend to describe time as unfolding linearly. We must be wary of anthropomorphizing reality, however. While our experience of time provides crucial clues, it is also important to situate that experience within the largest picture of reality we can imagine.

Consider the conceptual shift from a flat earth to a round one, and the enormous implications for exploration, trade routes, and map making. Similarly, situating our notion of linear time within a broader vision of a continuum between time and timelessness holds unexplored and undreamt possibilities. The second law locks us into one experience of time. A careful interpretation of the scientific concept of temporal symmetry offers an expansion of our idea of time and leads to many thought-provoking philosophical implications, as explored later. Through recognizing the limitations of both consciousness and the second law, we can release ourselves to explore the broader possibilities of reality.

The second philosophical limitation is in consciousness' inability to describe anything beyond its own experience. Do we perceive time as flowing in one direction because it does flow in one direction, or does it seems to flow in one direction because our brains perceive it that way. Imagine you are on a boat and you see an otter float up next to the boat from behind. Are you moving backward or is the otter moving toward you? Well, it depends on what you are measuring against, e.g., GPS coordinates, or from whose perspective you are measuring from, yours or the otter's. The answer changes depending on how many perspectives you want to include.

Kant draws the distinction between the world as we experience it through our senses, and the world in itself apart from our experience, preferring to focus on the former. He claims that we cannot know of the existence or non-existence of an object outside of space and time because it exists outside of our experience.[1] We can only speak of our own experience, always mediated through the forms of space and time. I suggest that we do experience the aspects of time beyond that of unidirectional flow, such as instances of reverse causality or timelessness, but that we can only communicate and conceptualize them from within time and space. When there are as many realities as there are observing subjects in the world then the task is not to discover which is "right," but how they fit together. A description of reality that encompasses alternative perspectives rather than pitting them against one another will surely prove itself superior.

By recognizing the limitations of the second law's scope of application to the statistical probability of manifest, isolated systems in terms of equilibrium states, we can then direct our attention to the potential broader perspectives of time available from the other realms of physics and subjective experience.

2.2   *Temporal Symmetry*

Recognizing these limitations, it seems natural to think of the second law as a special instance within the larger reality described by the laws of physics. Contrary to the second law of thermodynamics, most physical laws are time symmetrical. This means that they function equally well forwards and backwards in time. The time symmetrical equations include Newton's Laws, Hamilton's equations, Maxwell's equations, general relativity, Dirac's equation, and the Schrodinger's equation, covering classical mechanics, electromagnetism, relativity, and quantum mechanics.

While symmetry plays an important role in physics, asymmetry is equally important. Symmetry is balanced and static. Movement, however, requires some sort of imbalance, or asymmetry. Both symmetry and asymmetry are necessary to each other, symmetry for foundation and sustenance, asymmetry for change and novelty. While most of physics is time-symmetrical, the asymmetries present within quantum field theory and thermodynamics offer the tension that keeps things interesting. In cosmological evolution, symmetry breaking yields the emergence of the four fundamental forces, electro-magnetism, the weak force, the strong force, and

---

[1] (Tarnas 1991)

gravity. Might symmetry breaking play a role in the manifestation of time as well?

2.2.1 Forwards and Backwards, Simultaneously

What might temporal symmetry look or feel like? Most likely, it depends on your perspective—where in the flow of time you are standing, and which direction you face. We know that we typically experience unidirectional time. This suggests that temporal symmetry would include the possibility of backwards-flowing time.

Consider for a moment that you are standing outside of the flow of time. From this external perspective, the forwards and backwards flow of time would be coextensive and symmetrical. From within the flow of time, on the other hand, backwards and forward time look asymmetrical, with backwards time coming from the future towards the present moment and forward time coming from the past towards the present moment. However, if you look at the past coming toward you and the present flowing backwards into the past, then the two once again appear symmetrical. The symmetry of time is dependent on perspective.

Since we seem to experience the forward flow of time and not the backwards flow of time, we know there must be some asymmetries involved. Temporal asymmetry provides important clues but does not rule out the existence of its symmetrical counterpart, backward flowing time. The question, once again, becomes how does backwards time fit within our overall understanding of time?

The tricky thing about time symmetry is that we do not seem to experience backwards time. Our experience of time tends to align with entropy's arrow of time. In the same way that entropy may be a subset of a larger temporal reality, our subjective experience may too describe only a portion of a larger reality.

Often when people try to imagine time running backwards they imagine everything running backwards, such that they would experience growing younger, rain rising from puddles to clouds, or pulling food out of one's mouth and putting it back onto a plate, into a store, and eventually into the ground. Merlin, the wizard from the legends of King Arthur, lived backwards, amidst others' perceptions of forward flowing time. If someone did perceive time in such a way, it would not necessarily alter others' perception of the forward flow of time. Backwards flowing time may, in fact, be indistinguishable from forward flowing time because of its symmetry. Merlin provides an illustration of the simultaneous existence of backwards time with forwards time and posits the possibility of this alternative form of temporal perception. For this reason, I suggest calling the idea that backwards and forwards time occur simultaneously rather than mutually exclusively, a "Merlin" model of time. I suggest this model in order to entertain the notion that the backwards time of temporal symmetry would not necessarily be distinguishable from a forward time from our accustomed perspective of forward running consciousness.

This would, however, brings up larger questions of causality, determinism, and freewill. My concern here, however, is not to propose any sort of time travel or to propose a vast departure from the current realities of our experience, but to explore how backwards time already manifests itself. In upcoming sections, I discuss the limitations to our ability to interact explicitly with other realms of time.

One may then ask the point of entertaining such a notion. If backwards time is indistinguishable from forwards time, is the hypothesis impossible to test? The true test, however, may involve the question of whether or not such a notion may offer a perspective substantially different enough to reframe and explain some of the current challenges of physics.

Take for instance the quandaries of wave-particle duality represented by a photon interacting with itself in the double slit experiment, or entangled particles' action at a distance or faster than light particle interactions. If these particles are indeed participating in a realm of timelessness or reverse causality, our piddling objection to their lack of causal decorum seems irrelevant. Perhaps we could train ourselves, or may naturally evolve, to detect the subtleties of reverse causality, similar to the way we have evolved into our current understanding of time and extended our memory capacities.

2.2.2 *Quantum Temporal Oscillations*

Throughout this section, I will explore how and where backwards time shows up in physics. Several theories have emerged that suggest temporal reversibility has a role to play in realm of quantum mechanics. Feynman diagrams introduce this concept through their illustrations of particle interactions.

Beneath what we think of as empty space, hidden by Heisenberg's uncertainty principle, at the smallest scales of time and space, a seething quantum foam of virtual particles continually create and annihilate one another, just out of reach of detection. Imagine one such particle/antiparticle pair (as illustrated in the left hand picture of) simultaneously emerging from the quantum foam and then quickly snapping back together, annihilating each other before they can interact with any other particles.

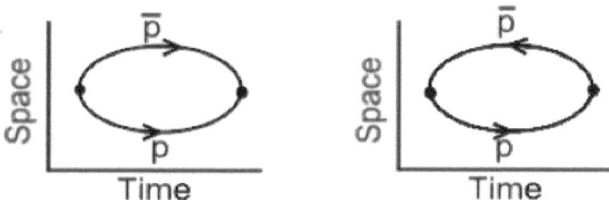

Figure 2.1 Particle/antiparticle creation/annihilation.
In the first image, at the dot on the left, a particle, p, and an antiparticle, $\bar{p}$, are created. At the dot on the right, they annihilate one another. In the second image, the particle emerges from the dot on the left forwards in time and turns around at the dot on the right to go backwards in time as an antiparticle until it turns back into a particle at the dot on the left.

Now imagine that the antiparticle moves backwards through time instead of forward (as illustrated above on the right). The point of creation becomes a point of temporal inflection, where the antiparticle, traveling backwards in time, turns around and becomes a particle traveling forward in time. The point of annihilation becomes another such inflection point. In this scenario, one particle oscillates between backwards and forwards time, appearing as particle and antiparticle.

In fact, Feynman diagrams often operate by the convention that an arrow pointing forward in time represents a particle while an arrow pointing backwards in time represents and antiparticle. This yields a vision akin to a standing wave, like a fast spinning jump rope, when viewed from a perspective of timelessness. Most physicists consider this as only a mathematical anomaly, not necessarily an indication that antiparticles actually travel backwards in time, but I think it worth considering. The Feynman diagram, or any diagram that makes time an axis, offers from an external timeless perspective from which to view it from.

Physicist John Wheeler proposed that perhaps only one electron bounces back and forth in time over and over again, creating the appearance of many, many electrons. Similarly, one photon may take on the appearance of many photons, because of their special relationship with timelessness, as discussed in a later section.

Another theory of forwards and backwards time in the quantum realm uses two state vectors to describe a quantum system, one that evolves forward in time and one that evolves backwards.[1] This theory acts in accordance with regular quantum theory and offers an objective description of EPR type issues, but has met with hesitancy because people do not know what to do with reverse causality.[2]

Do temporal oscillations matter on quantum level time scales? What effects emerge from there? Does it make sense to extend the quantum level oscillations between past and future to a macroscopic level? Does our picture of time become one of simultaneous oscillation between past and future rather than a unidirectional flow? Where else do we see evidence for possible temporal reversal?

Still honing our skills in traditional causality, our understanding of reverse causality might take some time to develop. The perspective of time as simultaneously oscillating between past and future, rather than flowing unidirectionally, re-frames our understanding of causality and may offer a more complete understanding of quantum theory.

2.2.3   *Spacetime Flexibility*

The idea of reversible time also has roots in the curvature of spacetime curvature as delineated by general relativity. Spacetime curvature shows up observationally through gravitational lensing. Gravitational lensing occurs when looking toward a massive gravitational object and an object directly behind it appears next to it. The hidden object can appear on two or even four sides of the foreground object. This occurs because the gravitational well created by the foreground object bends spacetime, so the path of light from the background object bends to go around the first object in order to take the shortest path to the observer. Light always takes the straightest spacetime path between two points.

---

[1] (Aharonov and Vaidman, 1990; Costa de Beauregard, 1989; Werbos, 1989)
[2] (Penrose, 1994, 389)

Time and Timelessness

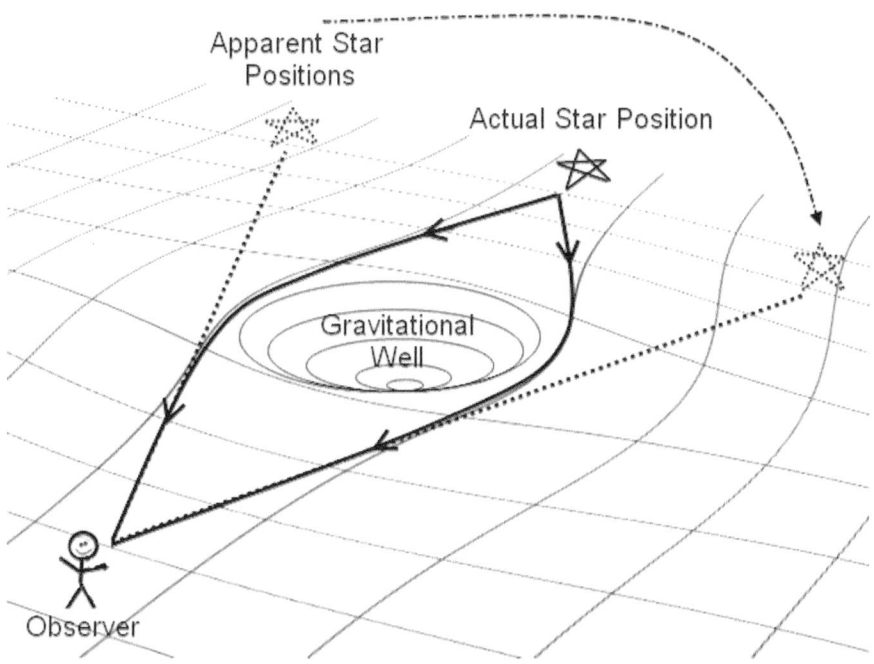

Figure 2.2 Gravitational Lensing

Note that a straight *spacetime* path is not necessarily a straight path through *space*, such as when curving around a gravitational obstacle. Sometimes the straightest spacetime path is around a gravitational source, rather than through the depths of its spacetime well. Think of airplane paths across the surface of a globe. The straightest path on a globe, appears as an excessively long curved path when flattened onto a two-dimensional map. On a flat map of a curved space, the straightest paths, or geodesics, reflect the curvature of the space. When the map reflects the curvature of the space, as a globe does, then the curvature of the straightest geodesics' curvature becomes evident.

Light cones can also help visualize this relationship between geodesics and curved spacetime. Light cones show all the possible photon paths from the past and into the future. A downward facing cone marks all the places a photon could have come from in the past. An upward facing cone maps all the places the photon can travel in the future. The vertices of both meet at the photon's present spacetime location. As a ratio of space per time, the speed of light defines the light cone slope. The light from everything we see travels along a light cone.

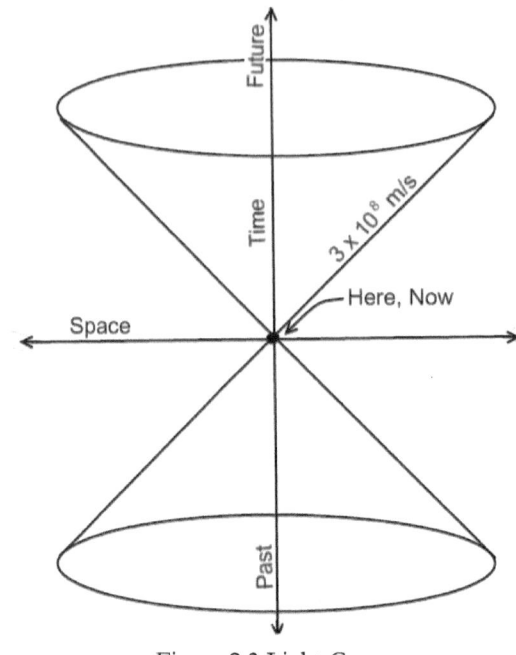

Figure 2.3 Light Cone

If something stands still, it still moves through time, so its path moves straight up along the time axis. If it is moving, it moves thorough space as well as time. Then its path slants away from the vertical path of a still object, adding some change in space to its change in time. These paths within the light cone demarcate timelike changes, because their unified spacetime path includes more movement through time than space. An object outside the light cone shares the same moment with the photon but in a different location. Objects outside the light cone differ from the photon by spacelike separations.

Along the light cone boundary photons traverse lightlike intervals, which are neither timelike nor spacelike, since the timelike separations and the spacelike separations cancel each other out, creating these null lines where photons cover 0 spacetime. The photon's spacetime path always measures 0. Photon are timeless in the sense that their paths do not involve timelike separation. Likewise, photons are spaceless in the sense that their paths do not involve spacelike separation.

Normally the light cone sits upright, but some coordinate systems describe light cones as tilting as they follow the curvature of spacetime as in the case of gravitational lensing. This can be explained as an artifact of the choice of coordinate systems but closed timelike loops, as discussed in the next section, are not as easily explained away.

2.2.4   Closed Timelike Loops

Light cones help describe the concept of closed timelike loops, a mathematically possible, but unobserved spacetime phenomena where traveling forward in time could land you in the past. If a light cone can tilt, then it can flip. If it can flip the future feeds back into the past.

Normally we think of time as progressing in one direction never to return to a point in its past. However, just as humans eventually discovered that if you travel far enough West you would eventually come back to where you started from the East, the same might be true for time. Just as the earth illustrates a closed spacelike loops, these might be a possibility for time as well. In a closed timelike loop, if you travel far enough, you'll end up where you started, in your past. Light cones illustrate this point, since the cones on opposite sides of the circle both lay in one another's past *and* future.

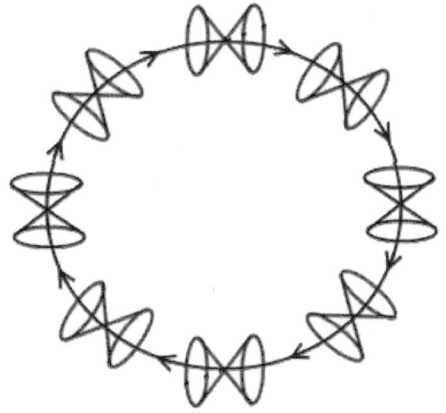

Figure 2.4 Closed timelike loop

One form of macroscopic time reversal appears in the closed timelike loops of Gödel's Universe. Einstein's field equations describe how matter and energy curve spacetime. Gödel's solution for Einstein's field equations, illustrated schematically in Figure 2.5, describes a rotating universe, which includes closed timelike loops. In Figure 2.5, notice the outermost circle forms a closed timelike loop, like that in Figure 2.4.

The second circle is a closed null curve, sometimes called a geodesic. A null geodesic is the path that light follows and the shortest distance between two points in spacetime. Null lines trace the edge of light cones and define their slope. In fact, this diagram simplifies the fact that every light cone has null lines spiraling toward and away from it along its edges making it the center of a torus of closed null lines. One could redraw the entire diagram at any point within itself by orienting it in accordance with the tilt of the light cone in that position.

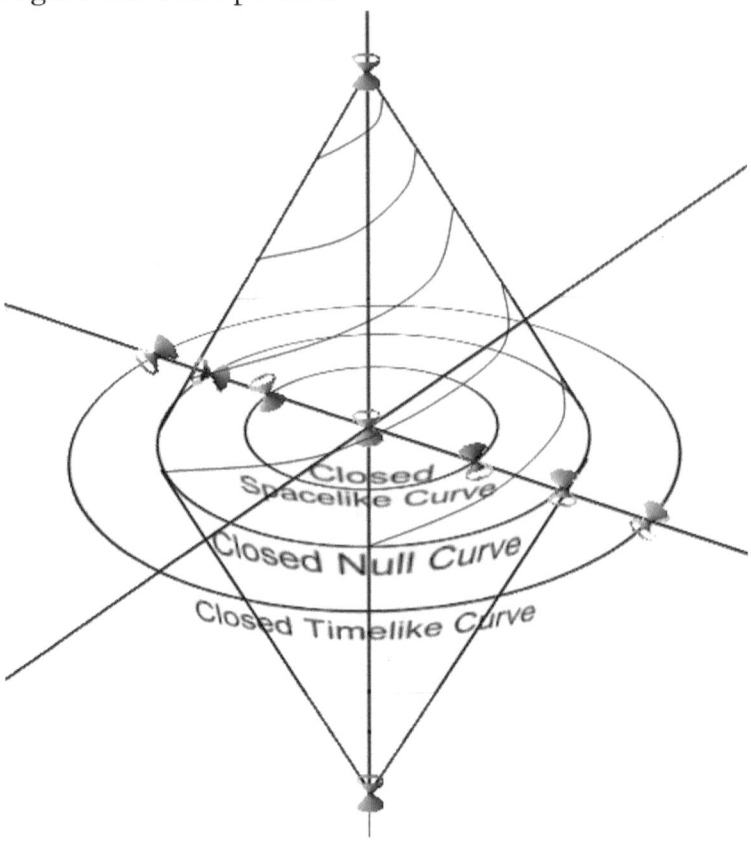

Figure 2.5 Gödel's Universe
(Adapted from Hawking and Ellis 1973)

Notice the small light cones that are tilted with respect to one another. The centermost light cone defines the vertical axis as time. The light cones on the outermost circle define the outermost circle as time forming a closed timelike curve. The middle circle emanates from the boundary of the light cone forming a closed lightlike, or null, curve. The innermost circle forms a closed spacelike loop, such as the surface of the earth.

In this same vein, the central axis turns into a closed timelike loop, just as the central axis of every light cone does. Each of these closed timelike loops exist at various angles to the one drawn in Figure 4.9. People often dismiss Gödel's universe as a universal model of spacetime because they do not observe a rotating universe. However, without an exterior reference frame what does rotation mean? In addition, in the spirit of our integral modeling, it might be possible that Gödel's universe describes smaller parts of a larger spacetime metric.[1]

Penrose speculates that closed timelike loops need not actually occur to have a real effect.[2] In quantum mechanics, just the possibility of an alternative can have a real effect of the outcome through its role as a counterfactual.

2.2.5   Counterfactual

The best example to describe the very real effects of the counterfactual is the Elitzur-Vaidman bomb-testing problem.[3] (I have taken the liberty of inserting the less violent trigger object of a firecracker in place of the traditional bomb.) Its explanation is somewhat involved, but conveys the power of the counterfactual, which allows us to determine whether some firecrackers would go off or not without actually setting them off.

The basic gist involves giving a photon two different possible paths to follow and seeing how it emerges when the paths reconverge. The key is that the photon emerges differently if one path holds a potential for measurement than with no possibility of measurement. With no potential for measurement the photon maintains superposition, "takes both paths," and interferes with itself when recombined. If a measurement device, like triggering a firecracker, is placed along one of the paths, the photon's path collapses into one path or the other. Collapsing into only one path yields a different outcome than maintaining superposition of both paths. Triggering the firecracker constitutes a measurement. Less obviously, if the firecracker is not triggered, but could have been, then we know the photon took the other path – also constituting a measurement, by the mere potential of measurement. There is also a possibility that the firecracker is a dud, in which

---

[1] (Hawking and Ellis 1973)
[2] (Penrose 1994, 259-270, 368)
[3] (Elitzur and Vaidman 1993)

case, we expected a measurement, but it was not taken, thus maintaining superposition.

If we get the result of two paths taken, we know the firecracker is a dud. If we get the result of one path taken, whether the firecracker explodes or not, then we know the firecracker is live. This shows that a possibility, even when not taken, still has repercussions. For those interested, the next section details the specifics of the Elitzur-Vaidman bomb-testing problem. For those less concerned with the details, feel free to skip the next section.

2.2.5.1 Elitzur-Vaidman bomb-testing problem

Consider a photon passing through a half-silvered mirror at a 45° angle to the path of the photon. Classically, it has a 50% chance of passing straight through and a 50% chance of being reflected at 90° to its original path. Quantum mechanically, according to the standard interpretation, the photon performs both actions and exists in a superposition of the two possibilities. If a measurement device collapses the superposition, it will detect the photon along one path or the other with equal probability just as the classical interpretation would assume. If, however, the photon interferes with itself before we take a measurement, the result proves the prediction of quantum wave interference and not the classical interpretation.

The wave interference results from a photon's phase shift of a quarter wavelength when reflected at a 90° angle. This shift is accounted for by multiplying the photon state by a factor i.[1] The photon in state A becomes a superposition of B and iC at the first half-silvered mirror. Then as B and iC round the bend, reflecting off full silvered mirrors they each gain additional i

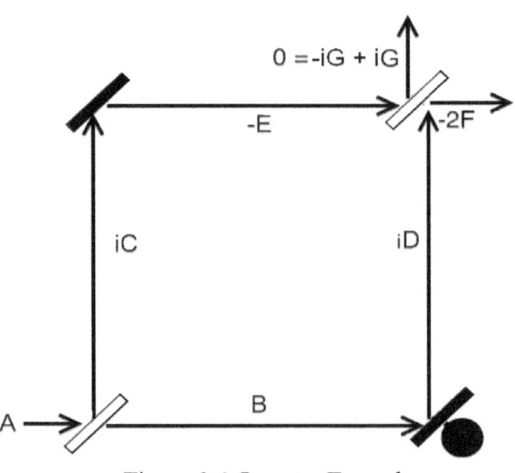

Figure 2.6 Counter Factual

---

[1] i is the square root of -1. So i^2 is -1. It is called a complex, or imaginary number. Multiplying by i is comparable to a quarter rotation around the complex plane, and thus as to a quarter of a wavelength. See: Wick Rotation.

factors, $B \to iD$ and $iC \to i^2 E = -E$, (since $i^2 = -1$). Then at the last half-silvered mirror, iD splits into a superposition of iG and $-F$, and $-E$ splits into $-F$ and $-iG$. $iG-F-F-iG$ now defines the entire system. The iGs cancel out and $-2F$ remains. This means that the photon will come out at F every time. There is no chance that it will come out at G.

Now we have a bunch of firecrackers, some of which are duds because their trigger is stuck. We place a firecracker behind the full silvered mirror on the lower right-hand side and make the mirror a little wobbly so if struck by a photon it will depress the firecracker's trigger and set it off. If the firecracker is a dud, the trigger is stuck in place, the mirror does not move, and the scenario plays out as described above with the photon exiting at F every time.

If the photon strikes the mirror in front of a live firecracker, however, the firecracker will go off. This action constitutes a measurement. Measurement amplifies the quantum system to register as a classical effect and so the superposition of states is broken and the photon travels only the $B \to iD$ path, re-entering superposition $iG-F$ at the last half-silvered mirror. When the firecracker is live (a measurement is taken), iG becomes a possible outcome. Therefore, we know that if the photon comes out at iG then the firecracker is live . . . which we already knew because the firecracker went off.

What if a photon emerges at iG and a firecracker did not go off? This can occur when a live firecracker is set up at the wobbly mirror, thus constituting a measuring device which collapses the wave function *whether it detects the photon or not*. A measurement taken at the full-silvered mirror yields a photon at B 50% of the time, setting off the firecracker, if live, and 50% of the time the photon will not register at B, meaning that the photon traveled the iC path. If the photon travels only the iC path, then, like the B path, there is a 50% chance that it will appear at F and a 50% chance that is will come out at G. However, when the photon takes the iC path, the firecracker does not explode.

Here is where the counterfactual comes in. The fact that the firecracker *could have* exploded affects the final path of the photon. If the firecracker could not have exploded, the photon could never reach iG. When the firecracker can explode, either it does or does not, and then the result is either F or iG. If the result is iG and the firecracker does not explode, we know that it could have, that it is not a dud, because if the firecracker is a dud, the result is always F. In this way, what does not happen tells us something definite about the state of the system. What is important here is that what *could have* happened gives a different effect than if it could not have happened, even if it does not happen.

2.2.5.2   Implications of the Counterfactual

In the Einstein-Podolsky-Rosen paradox, measuring one of two entangled particles instantaneously affects the reality of the other particle as well, suggesting faster than light information transmission, closed timelike loops, or global hidden variables. The Elitzur-Vaidman bomb-testing problem extends this mystery to suggest that the mere possibility measurement can affects the system in the same way a measurement might.

Thus, Penrose suggests that the mere possibility of closed timelike loops may cause reality to behave as if there were closed timelike loops, even if those loops do not actually occur. Penrose argues that closed timelike loops need only exist as counterfactuals to have real effects, thus avoiding the paradoxical implications of time travel, such as the possibility of somehow interfering with one's own birth, *a la* the 80's movie "Back to the Future."[1]

This suggests that the effects of any possibility have already "been considered," the effects of reverse causality are already incorporated into reality. One might expect the discovery of reverse causality to somehow change how reality plays out, but if the universe has always used it, or the possibility of it, then we just haven't noticed.

Considering the universe from a timeless perspective, as discussed in the next section, allows for overlapping backwards and forwards time. The interactions of these possibilities might impose certain limitations for information transmission across time in either direction, preventing time-travel paradoxes.

Past information is compressed and carried into the present through memories and through cause and effect. Its compressed form often renders aspects of past information inaccessible in the present. For example, you may carry memories that you cannot bring up at will, but that you know you carry because they may arise spontaneously in response to a specific smell or sound. Like the methods of information retention or transmission from past to present, perhaps the transmission of information from future to

---

[1] (Penrose, 1994, 383)

present, may face similar obstacles, such that we are normally unaware of this direction of information flow.

Thus, if closed timelike loops occur, the process of traversing the closed timelike loop might compromise the completeness of information passage, preventing our awareness of the loops. The next section also discusses this possible loss of information in relation to the bending of time and time dilation.

2.3 The Expanded Moment

2.3.1 The Speed of Time

Reverse causality offers an important perspective but gains even more depth through the perspective of timelessness. Despite Penrose's work with null lines and emphasis on the relationship between consciousness and timelessness through Plato's realm of ideas, Penrose does not appear to address what seems to be the natural combination of these concepts, the timelessness of the photon.

Timelessness offers a counterpoint to the linear flow of time. The space between the time and timelessness opens up into a continuum differentiated by velocity. On one end of the continuum a still object moves through time without moving through space and tracks time at the same rate as other still observers. On the other end of the spectrum, the speed of light presents a cliff in spacetime, where as one speeds up to approach it, time slows down and space contracts, suggesting the approach of a timeless point--the photon, unreachable by speed, yet so intimately intertwined in all of matter's finite space-time activities.

Objects moving at the speed of light, e.g., photons, appear to us, material beings, to move through both time and space. The photon path, however, by virtue of traversing lightlike intervals, does not traverse spacelike or timelike separations. At the speed of light, space and time cancel each other out.

Movement in space and time detract from one another. When in stillness, one moves only through time, rather than space. As one gains speed, movement through space detracts from movement through time. At the speed of light, space and time's opposition cancels each other out.

If photons traverse no changes in time or space, then for them all space and time exist in one point. Only from a slowed down perspective, which extends space and time, do photons appear multiplicitous and distributed. Although photons may seem to us to traverse all of spacetime, they, by definition, traverse no spacetime.

From an external perspective, something moving at the speed of light moves through external space and time but appears internally indivisible by time or space. A stationary observer cannot perceive any temporal flow or spatial extension within the interior of a photon.

Special relativity describes that the faster something goes the slower time goes for that object, although the disparity only shows up in comparison to a slower reference frame. Time always feels the same within your reference frame, no matter how fast you are going.

Time dilation describes the phenomenon of time moving slower as an object's velocity increases. The mathematical representation looks like this:

$$\Delta t'/\Delta t = \sqrt{1-(v/c)^2}$$

Such that, an observer who experiences a time change, Δt, sees the time change of a second observer, who is moving at a velocity, v, as $\Delta t'$. c is the speed of light.

Using this equation, we can see just how time changes with velocity. The equation defines the ratio of $\Delta t'$ to Δt. The "speed of time" is this ratio between two different rates of temporal passage. If the one observer experiences the passage of one second as equal to another observer's experience of two seconds when they come back together, then the first observer's time has dilated–their moment has expanded.

We have evidence for time dilation in the slowing of clocks transported in very long flights and the extended lifetimes of particles traveling at speeds close to that of light.

What does it mean for time to slow down? On one hand, the dilation of time seems to indicate a swelling of the moment, encompassing more of the past and future than one has access to at slower speeds. In spatial terms, a moving object covers larger amounts of space in smaller amounts of time. From a car traveling at 60 mph, you can see a whole town pass by in 6 minutes.

You may see a whole town pass by in 6 minutes when traveling at 60 mph, but you only see one moment of that town, a cross section in time, and you only see the surface. The realm of interaction may have expanded, but the actual ability to interact is inhibited. To see the backside, the inside, the minute details, the rhythms and patterns of the town requires a closer synchronization of time scales. To know something that is still, make yourself still. To know something that is moving, follow it.

The same may be true for time dilation—increased range of temporal overlap, decreased potential interaction with other times. The limitations of communication between time and timelessness offer important insights into the limitations of time travel and precognition.

2.3.2   The Frozen Timeline of the Block Universe

Special Relativity spatialized time by multiplying it by velocity. Time, when multiplied by distance per time, cancels out, leaving distance. Using this convention, mathematician Hermann Minkowski added time to create a 4-D version of the Pythagorean theorem, formulating the block universe, also known as Minkowski space.

The block universe spatializes time, imagining all time as a static landscape. This static landscape, however, is four dimensional, merging timeline and map. The inclusion of all time yields a static, spatialized picture of time rather than the sense of temporal flow typical of human experience from within a slice of time.

Our mathematics can describe a four-dimensional reality and we seem to live within a four-dimensional reality, yet our imaginations stretch to attempt to visualize four dimensions, when our experience only allows us to see three. The dimension of time remains rather invisible, revealing itself one slice at a time, and then vanishing into interior dimensions of memory.

The timeline approach to visualizing the block universe involves imagining everything smeared out through time--each moment slightly displaced from, though still connected to the previous moment. In this version we represent time as space. Relativity's mathematics spatializes time in order to bring it into dialog with space.

Consider an alternative vision, with time as layers in space, instead of spatial displacement. In this version the stationary world stays pretty much the same instead of shifting in space to show a change in time, while moving objects appear as transparent smears through the stationary space. Your regular three-dimensional body smears out across all of the places you have visited over your lifetime, like a photograph taken with the shutter left open for a long time, collecting all your moments in one continuous chain.

You might appear more solid in places where you spend a great deal of time and more transparent, or ghost like, in places where you just passed through once. A large amount of time in one place might appear with greater solidity or density because it contains many layers. This begins to get at time as a measure of density or depth, rather than length. All beings from all times would take on this smeared-out, layered look, all overlapping. It becomes too dense to unravel, and one might begin to feel grateful we experience a limited slice of time, instead of all of time at once.

Spatializing time unravels this density. Imagining the fourth temporal dimension of time as a spatial dimension, spreads that density out over another dimension. Then even standing still smears out into a worm or tube through the dimension of time.

Special relativity spatialized time, t, in the mathematical description of the universe by multiplying it by the speed of light, c. When multiplying time by speed (distance per time) time cancels out, leaving only a measure of distance, like the other spatial dimensions.

Minkowski's block universe provides a geometrical description of the universe, like the Pythagorean theorem ($a^2 + b^2 = c^2$), except in four dimensions and with time multiplied by the speed of light and subtracted instead of added. Making space and time opposite signs preserves the spacetime invariance under special relativity's Lorentz transformations.[1] Subtracting the square of $c\Delta t$ however yields a measure of imaginary time, as the square root of -1 = i, an imaginary number, which further distinguishes time from space, as we will explore later. The complimentary relationship of space and time in special relativity is such that moving through space faster means you move through less time. Hence the opposites signs of space and time.

$$\Delta s^2 = \Delta x^2 + \Delta y^2 + \Delta z^2 - (c\Delta t)^2$$

Δs is the change in proper time as measured by a local clock, which if accelerated tends to run slow. In contrast, Δt is the change in coordinate time, as measured by a distant observer unaffected by local acceleration. This Einstein-Minkowski block universe metric, though it does include two measures of time, is essentially timeless because it includes the expanse of time without the progress of time. The universe, including its history and

---

[1] (Nahin, 1998/2007)

future, can be laid out in this combination of map and timeline.

Objects take up a specific amount of space-time, the total of which is recorded by Δs, proper time. Δs gets divvied up between movement through time and movement through space. By sitting still, one moves through time but not space. Moving through space, detracts from an object's motion through time. Light moves so quickly through space that its change in space equals its change in time, canceling each other out and leaving zero change in proper time. For light, proper time, s, equals zero.

This static picture of all-time contrasts with our experience of past, present, and future as distinct from within an individual moment of time. This contrast however does not render these two versions of time incompatible, but rather reveals the distinction of the perspectives, despite the invariance in their spacetime measurements. The existence of a timeless perspective does not rule out an experience of temporal flow, or its reversal for that matter.

2.3.3 *Temporal Symmetry as a Subset of Timelessness*

In this section, I propose temporal symmetry as a subset of timelessness and specify the relationship between both concepts. Time is a subset of temporal symmetry, which in turn is a subset of timelessness. Timelessness and temporal symmetry are distinct from time in that they do not obey the unidirectional causality of time. Reverse causality may simply appear as acausality to those rooted in a unidirectional flow of time, even though reverse causality may have its own logic. Timelessness and temporal symmetry are distinct from one another in that temporal symmetry requires that the transition between points in time proceed linearly through connected moments in time, whereas timelessness suggests a simultaneity and unification of temporal moments. This distinction is similar to the distinction between retracing your steps to try to remember where you put your keys and remembering a childhood experience that does not require you to remember every event between now and then.

Since we do not yet perceive reverse causality, from within time we can only distinguish between two categories causality and acausality. A more nuanced perspective might eventually determine whether something we perceive as acausal is an effect of reverse causality or of timelessness. Since temporal symmetry allows motion both forwards and backwards in time it can access any point in time in the same way timelessness can, whereas unidirectional time restricts one's access to only the present and the future. Thus, due to the limited perspective from within time, it is difficult to distinguish between the effects of timelessness and temporal symmetry. A premonition not to get on a plane that later crashed could be an effect of reverse causality or of tapping into a timeless knowing. The mechanisms of timelessness and temporal reversal may differ, but these are invisible from within time. From a perspective bound by time, temporal reversibility and timelessness are easily conflated as they both manifest in the same apparently acausal way.

Bohm provides a great example of this overlap of temporal reversibility and timelessness using the quantum time symmetrical oscillations of the vacuum state.

> in modern quantum-mechanical field theory the "vacuum state" has, properly speaking, no physically meaningful notion of time at all in it. Or, more accurately, in the vacuum state the "state-function" (which represents the whole of space and time) oscillates uniformly at a frequency so high that it is utterly beyond any known physical interpretation. . . If time has to be abstracted from an ordered sequence of changes in an actual physical process, we would be justified in saying that the vacuum state is, in a certain sense, "timelessness" or "beyond time."[1]

In this sense, temporal symmetry is a form of timelessness occurring at very small scales of time and space where energy and frequency tend toward infinity.

2.3.4   *Time and Timelessness*

Having discussed quantum and cosmological temporal reversibility as a form of timelessness, let's focus on the relationship between time and timelessness.

On one hand, time and timelessness are not separate from each other; they are two sides of the same coin, similar to the relationship between matter and energy. David Bohm's notion of an implicate and explicate order corresponds to the realms of timelessness and time,

---

[1] (Bohm 1986, 189)

respectively, and exhibit this mutual enfolding of one another, similar to the boundary of the fractal Mandelbrot set. Bohm states,

> each system has to be seen in both aspects, i.e., of time and of a relatively 'timeless' enfolded state. In the time aspect is comprehended the *becoming of being*, while in the 'timeless' aspect is comprehended the *being of becoming*.[1]

On the other hand, while time and timelessness are always present within each other, our experience of them is limited. The choice to participate in one, limits participation in the other. Like continuous variables, they somehow limit one another. Position and momentum, in the uncertainty principle, are good examples of continuous variables. The more accurately you know one variable, the less accurately you know the other. In the case of relativity, the more space you cover per unit of time, the less time you traverse compared to stationary time. The more you participate in timelessness, the less you participate in time.

We know this sort of relationship intuitively through many everyday dualities. Breadth and depth appear as continuous variables in the expression, "Jack of all trade, master of none." This principle also alludes to the limitations inherent in interdisciplinary explorations such as this one. Both of these examples stem from the limitations imposed by time. Breadth and depth both require time, and one can only pursue one at a time. Time is the interiority that fleshes out space. There seems to be an inherent tension between space and time, time and timelessness, breadth and depth, transcendence and immanence, similar to the tension between continuous variables. In fact, it seems likely that many of the known sets of continuous variables, such as those of the uncertainty principle, stem from the underlying tension between space and time, time and timelessness.

The way to access other points in time, without having to go through the entire linear progression to get there, is to step out of the line in order to drop in at another point on it. Often in mathematics, the way to solve a problem is to add another dimension. Stepping out of time is stepping into timelessness.

Consider the timelessness of the photon. Since a photon exists outside of time, it also exists outside of temporal distinctions, meaning here and now is no different from 10 minutes from now. The photon experiences

---

[1] (Bohm 1986, 197)

them simultaneously from outside of time. Bohm captures this relationship between the timelessness of the photon and time perfectly in his description of the internal eternity and external motion of the now, though he does not mention the role of the photon in this dynamic.

> The deeper that consciousness goes into the implicate order, the more similar these moments will then be and the less significant will be their differences. If it were possible for consciousness somehow to reach a very deep level, for example, that of pre-space or beyond, then all 'nows' would not only be similar–they would all be one and essentially the same. One could say that in its inward depths now *is* eternity, while in its outward features each 'now' is different from the others... There is a well-known saying that 'now' is the intersection of eternity and time. On its inward side, 'now' is, as we have seen, ultimately the same as eternity. On its outward side it participates in movement.[1]

It is this possibility of voyaging into this timeless internal eternity that offers the possibility of contacting temporal moments external to the now.

2.3.5 Subtle Causality

Stuart Hameroff neatly ties the theory of forwards and backwards time to the scale limitations of entropy, EPR quantum entanglement, and subjective experience via he and Penrose's theory of quantum consciousness. He states,

> In the quantum realm, time is uncertain, and events may run "backwards." The second law of thermodynamics apparently does not apply in quantum systems. Quantum state reductions such as OR events may send quantum information "backwards in time" . . . Backwards time referral of quantum information can account for effects in EPR entanglement . . . it is quantum information rather than classical information that travels backwards. In Orch OR quantum information is "pre-conscious" or "sub-conscious,"

---

[1] (Bohm 1986, 199)

becoming conscious at the "Now" moment of objective reduction. Thus, each conscious moment incorporates quantum information from the past and the future . . . [1]

If indeed each quantum collapse and each conscious moment incorporate quantum information from the future, then this future information, hidden to us, could be the hidden variable, which makes quantum systems non-deterministic. If the variable we need in order to determine the state of a quantum system resides in the future, we will not be able to determine the state of that system until we figure out some way to access the future. There are in fact studies which seem to physiologically record just such a future influence. Though people do occasionally seem to have extraordinary contact with the future, either as psychics or through divinatory practices such as the I Ching or astrology, the accuracy of the information they bring back seems limited by probability as well because it often arrives in the guise of archetypal patterns, the symbolic language of the unconscious. There are in fact studies which seem to physiologically record just such a future influence. The fact that our bodies may register the influence of the future, just below our conscious awareness, offers a physical correlate to the abilities of our unconscious.

In two experiments, Dick Bierman with Dean Radin, and later with Steven Scholte[2] used fMRI brain imaging and took skin conductance measurements on participants viewing carefully randomized images of either neutral or emotionally charged subject matter (i.e., of a violent or erotic nature). The subjects viewed a blank screen for several seconds. Then an either neutral or emotionally charged image appeared for several seconds. The blank screen then returned for a longer period before the next round began. Instruments measured the volunteers' physiological responses the entire time and showed the expected peak and decline after the image viewing. Remarkably, a smaller peak also appeared up to four seconds *prior* to the display of an emotionally charged image, though not with neutral images.[3] They termed this phenomena "presponse." This is exactly what one might expect to see as evidence for participation of consciousness in the timelessness of an expanded moment and/or via sub-conscious/quantum reverse causality.

---

[1] (Buccheri et al. 2003, 86)
[2] (Bierman 2002), (Bierman and Scholte 2002),
[3] (Bierman 2002, 9; Bierman and Scholte 2002, 7)

Penrose addresses phenomena such as this, including experiments that suggest that we make decisions before we are aware of making the decision[1] and that we perceive sensation as simultaneous with the stimulus rather that at a slight time delay as one would suspect to allow for nervous system transmission. "The mildest possibility in accordance with this would be a non-local spreading time, so that there would be an inherent fuzziness about the relationship between conscious experience and physical time."[2] This also corresponds neatly to the expanded moment and to quantum temporal reversibility.

The perspectives of temporal reversibility, closed timelike loops, and timelessness offer possible explanations for EPR type paradoxes and a seat for consciousness's interaction with timelessness. What appears, in quantum mechanics, to be superluminal communication or non-local action at a distance, makes much more sense with a flexible framework of time.

As discussed in section 1.3.4, the EPR paradox illustrates that measuring one of two, separate but entangled, particles somehow affects the other particle instantaneously. The paradox arises because the assumption of unidirectional causality dictates that either the communication between the particles occurs faster than the speed of light or the continuity of space (locality) is somehow violated.

As discussed throughout this chapter, however, temporal reversibility and timelessness offer mechanisms to connect disparate points in time and space. For one, if the particles are photons, they are moving at the speed of light, which, according to special relativity, means they exist in a state of timelessness, and may even be the same particle. Timelessness steps outside the linear continuity of time and space, collapsing temporal separation and makes faster than light messages unnecessary or at least redefines their mechanism. In addition, if the two entangled particles were in fact the same particle, simply appearing as two particles from a temporal perspective, then one particle would obviously affect the other instantaneously.

Timelessness offers an additional dimension in which closed timelike loops can participate. The acausality of timelessness or closed timelike loops could offer the additional dimension from which consciousness can reflect on itself and provide the mechanism for the basis of consciousness' unique abilities.

---

[1] (Keim 2008)
[2] (Penrose, 1994, 387).

3
Fractal Time

Looking at time and timelessness in the context of quantum consciousness and relativity, sets the stage for the entrance of fractals as a model of time to relate consciousness and cosmology. This chapter explores five fractal properties that speak to properties of time, building a fractal model of time. Utilizing the fractal spacetime of Laurent Nottale, and Susie Vrobel's fractal model of subjective time, these shared properties of fractals and time begin to emerge. The five fractals properties that reverberate throughout this chapter include: self-similarity, nature's patterns, embedded dimensionality, the infinite within the finite, and the role of imaginary time.

This chapter utilizes two existing fractal time models to explore how fractals interface with time. Laurent Nottale's model of fractal spacetime provides key insights into the usefulness of fractals and links the fractal model to the quantum temporal reversibility discussed in previous chapters. Nottale does not explore how one might experience a fractal model of time subjectively, nor does he address the issue of timelessness. Susie Vrobel, however, does map a fractal model of time onto subjective experience, specifically linking fractal primes to timelessness via Penrose's concept of insight. While Vrobel's model is philosophically satisfying in ways Nottale's misses, it does not develop the mathematical connections as thoroughly as Nottale's rigorous formulation. Thus, by bringing Nottale and Vrobel into dialog, I propose a more complete extension of them both, and draw attention to several more specific links between a fractal model and subjective experience of time.

This book uses the kernel of a fractal model of time to begin to understand how consciousness interweaves with the temporal fabric of quantum and cosmological reality, and in the process, it shifts our understanding of the physical universe and ourselves. We have typically separated consciousness from external reality in order to describe each of them more simply. Time, however, is integral to both. To get a clear picture of what time is we must reintegrate our understandings of consciousness and external reality. Using the models of Penrose and Hameroff, Vrobel, and Nottale, I strengthen the link between fractals and time, and refine the connections between the quantitative and qualitative boldly begun by Vrobel.

3.1 Fractals

What is a fractal? A fractal contains multiple scales of irregularity. Sometimes that irregularity is self-similar, generated by the reiteration of a mathematical formula. Reiteration is a form of feedback, where the answer to the formula recycles into the original formula to generate the next solution. Each round is an iteration of the formula. The Mandelbrot set example, in the following section, illustrates this principle. Like mathematical reiteration, time's iterative cycles demarcate temporal progress—repeating cycles of years, of seasons, of days, of lifetimes, of historical patterns.

Self-similarity means that the pattern of the whole repeats within each of its parts. The smaller copies of the pattern nest within the whole pattern. The microcosm reflects the macrocosm. Think of Russian nesting dolls. The concept of nesting in fractals is similar, except that instead of there being only one doll of each size, fractal patterns tend to multiply as their scale decreases, more like a family tree with lots of kids in each generation.

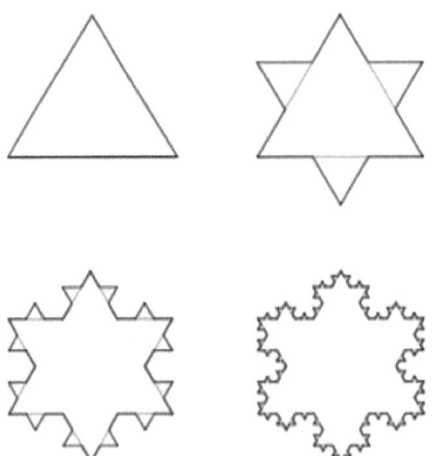

Figure 3.1 Koch Snowflake

Self-similarity gives fractals their characteristic irregular structure. Zooming in to try to find a straight, predictable line often reveals more detailed irregularity. The fractal, or Hausdorff, dimension defines the fractal's complexity or density.

Take the Koch (pr. k-aw-kh) snowflake for example. Imagine the Star of David, made of two triangles pointing in opposite directions. Now take one of the smaller triangular points of the star and turn it into a Star of David by adding another triangle pointing in the opposite direction. This process can continue infinitely, generating an infinite perimeter within a finite space.

With irregularity across scales, fractals provide a mathematical context in which infinity within a finite boundary makes sense. "A fractal is a way of seeing infinity."[1] Time always runs into the notion of infinity, whether through the notion of eternity or the timeless depths of the ever-present moment. Thus, fractals may offer a way to visualize the infinite nature of time as experienced within the finite bounds of a moment.

In the Koch snowflake example, the fractal object emerges through the reiteration of a simple procedure. The fractal's self-similarity also manifests as the shape of the whole repeats at various intervals of scale. Holographic film exhibits a similar quality. Shining light through the entire piece of film yields the most accurate image; whereas cutting the film into small pieces and shining a shining a light through one piece of the film also display the entire image, only at a decreased resolution. The entire image is stored in each of the parts. Fractals also repeat the pattern of the whole within their parts, but without the loss of clarity that occurs in holograms.

---

[1] (Gleick 1987, 98)

3.1.1 Mandelbrot Set

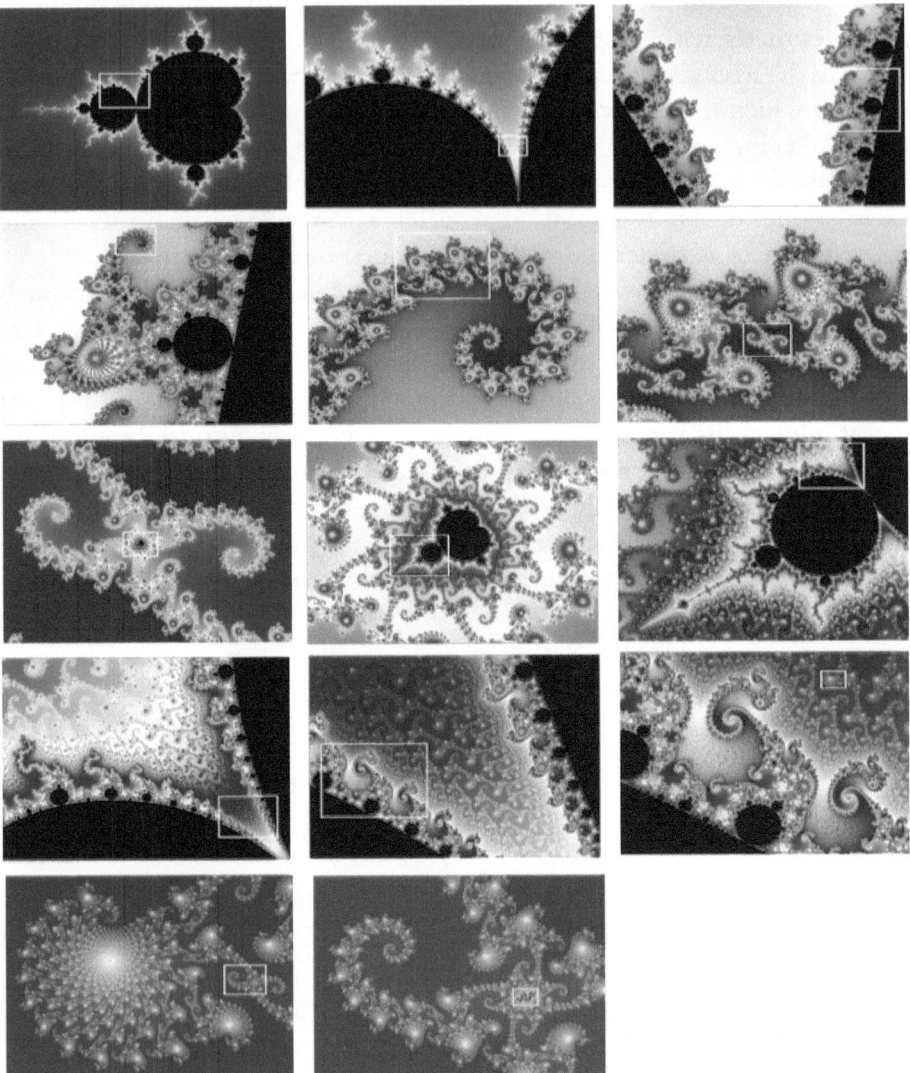

Figure 3.2 Mandelbrot Set
Starting at the upper left each successive illustration shows an expanded portion of the one preceding it.[1]

---

[1] (Beyer 2019). CC BY-SA 3.0

A FRACTAL TOPOLOGY OF TIME

Like many instances of insight, mathematician Benoit Mandelbrot sees his discovery of the Mandelbrot set as less of an act of creation, than a discovery of a pre-existing mathematical reality, just waiting for consciousness to stumble upon it.

Zooming into the Mandelbrot set, complexity unfurls. Spiraling tendrils, psychedelic zippers, and fuzzy Buddha bulbs reappear again and again across vast differences in scale, yet without exact repetition. One may expect to find a "bottom," or lowest limiting scale, but it keeps going, limited only by available computing power. The Mandelbrot set, illustrates the fractal paradox of infinite depth within a finite area, opening a powerful parallel to the paradox of the present moment's eternal, yet ephemeral nature, offering a visual illustration of the intertwining of infinity and finitude, timelessness and time.

The Mandelbrot set colors each point on the complex plane by plugging it into a reiterative equation and seeing what comes out. How many iterations of the algorithm it takes to escape a specific radius ($1+ \sqrt{2} \approx 2.414...$) determines the color. A point that does not escape the designated region after many iterations, is colored black and considered bounded, or more accurately, undefined, as its definition (as bounded or unbounded) requires too many iterations to accomplish. The algorithm consists of squaring the initial point, c, and adding it to itself, then doing the same for the solution.

$$c \rightarrow c^2 + c = c_2 \rightarrow c_2^2 + c_2 = c_3 \rightarrow c_3^2 + c_3 = c_4 \rightarrow ...$$

One can never know for sure whether a given point in the black area will ever escape the radius, because we cannot perform an infinite number of iterations. Thus, the black area, the Mandelbrot set itself, is non-computable. This renders the edge between the black and colored portions as also undefinable. As one seeks the boundary, zooming in to find its definiteness, the boundary continues to twist and turn, refusing definition.

The infinite (or at least, indeterminable) complexity of the boundary renders the Mandelbrot set non-computable. The more you try to distinguish the black from the colored points, the more they intertwine--a fundamental principle of duality in any realm.

The depth of scale allows for an infinitely long line to exist within a finite surface area, by virtue of finer and finer crinkliness. The Mandelbrot set itself, as the boundary line between the black and colored points, has a topological dimension of one (the boundary line), but a fractal dimension of two--known as a space filling curve--because the boundary turns back on itself so much that it completely fills the two-dimensional space. In the same way a 2-D flat piece of paper, when crumpled into a ball, takes up more 3-D space.

British mathematician Lewis Fry Richardson hoped to correlate the lengths of boundary between two countries, with their probability of going to war. In the process he discovered significant discrepancies in the measured lengths of the boundaries. He recognized that different sized measuring sticks produce different lengths for the boundaries. Benoit Mandelbrot popularized this paradox as the "Coastline paradox."

Mandelbrot challenged anyone to measure the coastline of Britain. "No problem," one thinks at first. But then people come back with as many different measurements as there are measurers. Those using shorter measuring sticks return longer measurements. Why?

Using a kilometer-long measuring stick ignores certain nooks and crannies along the coast. Using a meter-long stick accounts for more irregularities, resulting in a longer overall measurement than the kilometer-long stick measurement. The length of the coastline further increases by employing a centimeter-long measuring device, thus incorporating even more irregularities.

As the measuring stick gets smaller, the length gets longer. This reasoning process then concludes that the most accurate measurement uses an infinitely small measuring device, yields an infinite coastline length. Yet the island takes up a finite space. Herein lies the paradox.

This same paradox seems to apply to our experience of time. Likewise, if we consider the present moment as eternal, yet divided and deepened by repetition, the depths of scale begin to speak to the depths of time. Measurement in years, months, or minutes may all add up to the same length of time, but do those count each emotional rise and fall, or the complexity or barrenness of each moment? Do our temporal measures gloss over the infinite depths present in each moment? Infinity is not easy to work with. Physicists and mathematicians have long chosen to ignore the true depths of such mathematical monsters.

Penrose brings in the fractal model's non-computability as a possible mathematical mechanism that might allow consciousness's contact with timelessness. I suggest taking Penrose one-step further and aligning the Mandelbrot set itself, the black, bounded, non-computable portion, with timelessness. The Mandelbrot set's non-computability offers a possible mathematical illustration of the potentiality of timelessness. Time emerges from timelessness, as explicate (manifest) reality emerges from (potential) implicate reality. Likewise, the colored regions emerge from the indefinable blackness, unfurling from the receding interface between the two, like time from timelessness continually in dialog, every moment in dialog with the larger expanse of time.

Using the example of the Mandelbrot set, we can begin to consider other fractals properties that might model temporal mysteries. Fractal dimensionality offers way to quantify the density of the depth within a finite boundary.

3.1.2   Fractal Dimension

Typically, dimension refers to the magnitude, or extent, of something in a particular direction, such as one of the three Cartesian coordinate--x, y, z--height, length, or width.

We could not easily tell the difference between the words on this two-dimensional (2-D) page if we did not exist in an additional third dimension. The perspective offered by an additional dimension reveals not only the divisions within the lower dimension, but also the unity that holds them together. For example, by transcending the two-dimensional surface of a page, we recognize not only the distinction between words on that page, but the commonality of the page in which they exist as well.

In mathematics a problem, unsolvable in a lower number of dimensions, may find resolution through adding a dimension. Adding a dimension offers transcendence, going beyond previous limitations. Consider the Pythagorean theorem for example. The relationship between the 1-D length of the sides, in a particular 2-D arrangement, is actually a relationship between 2-D areas or "squares." By considering each of

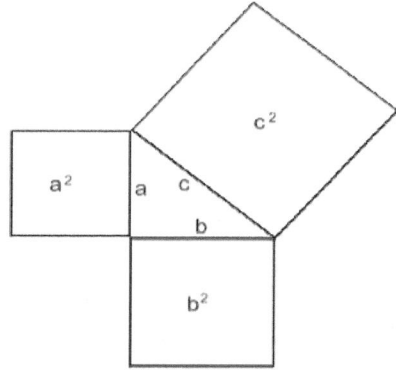

Figure 3.3 Pythagorean theorem:
$a^2 + b^2 = c^2$

the lengths as a plane instead of a line, an unknown length of the triangle can be determined. This illustrates how an additional dimension can reveal relationships in lower dimensions.

Fractals have a unique relationship to dimensionality. Fractal dimensions are not restricted to integer dimensions, such as one or two, but include fractions of whole dimensions, such as 1.7. An object's fractal dimension exceeds its topological dimension by a fraction of the next available dimension, determined by how much of the next dimension it fills in. The topological dimension is equivalent to what we normally think of as dimension--a 1-D line, and 2-D square, and 3-D cube. A curvy line has a topological dimension of one, but requires two dimensions in which to curve, thus taking up a portion of 2-D space.

The Hausdorff dimension of the Koch snowflake for example, is 1.26. It is more than a one-dimensional line, but not quite enough to constitute an entire solid two-dimensional plane. A 1-D line's topological dimension, however, is still one. The Mandelbrot set on the other hand, is a space-filling curve, the boundary line between black and color is everywhere if you zoom in deep enough. The 1-D boundary of the Mandelbrot set takes up the entire two-dimensional plane and thus has a fractal dimension of two.

The more the fractal fills in the two-dimensional plane, the closer its fractal dimension will be to two. "Fractional dimensions become a way of measuring qualities that otherwise have no clear definition: the degree of roughness or brokenness or irregularity of an object."[1] This fractal dimension corresponds to the object's density, depth of scale, and complexity. While not enough to increase the topology by an additional dimension, the fractal dimension offers it texture.

The Hausdorff dimension's equation is a ratio between a pattern's dilation and multiplication across scales,

$$D = \log n / \log s$$

where n is the number of multiples of the original pattern at the next smaller level. The scaling factor, s, is the number by which the pattern is divided by to get to the next smaller version of itself.

---

[1] (Gleick 1987, 98)

Fractal Time

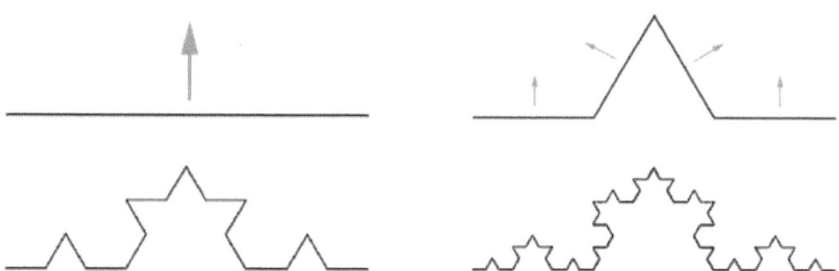

Continuing with the Koch example, consider just one side of the triangle. Divide one side by three and bump the middle section up into a peak. Each new line segment is a third of the original line and there are four copies. The new line is 4/3 times the length of the previous line--the original length divided by three and multiplied by four. Dividing the original line by three yields a scaling factor, *s*, of 3. There are now four copies of one third of the original line, so the number of self-similar structures per scale, *n*, is four.

$$D = \log 4 / \log 3 = 1.26$$

The dimension is the same at every level of scale in a fractal. Now, with a basic understanding of some fractal properties, we can begin to consider the possibility of time as a fractal phenomenon.

3.1.3   *Fractal Models of Time*

What is a fractal model of time? Fractals seem to offer a promising mathematical model for time, because many of their properties parallel properties of time. In both nested cycles build unpredictability and infinite depth within a finite range. For example, both fractals and time have unpredictable rates of change. The way we experience time as progressing slower, faster, or as standing still is similar to the unpredictable rate of change, or non-differentiability, of a fractal function. In addition, repeated and nested cycles build both fractals and time. Time is built by the cycles of the clock, the earth's spinning, the moons phases, and the seasons, etc. Fractals are built by the reiteration of a foundational equation or driving process. Fractals and time both contain complex relationships between infinity and finiteness. Similar to a fractal's paradox of infinite length and finite surface area, the moment is eternal, yet also passing away. The moment contains all experience, both past and future, yet is also bounded by the past and future. These, among other properties, make fractals well suited for describing our temporal experience.

This chapter explores these properties in greater depth as well as looking at two other models of fractal time, one from the perspective of physics, Laurent Nottale's, and one from the perspective of subjective time, Susie Vrobel's.

3.2 Nottale's Fractal Spacetime

Penrose emphasizes the need for a new worldview that could provide, for quantum mechanics, the additional depth that general relativity provided for classical physics.[1] General relativity included and expanded beyond classical mechanics by giving up the assumption of straightness and allowing spacetime to curve. French astrophysicist Laurent Nottale postulates a more general theory that subsumes both relativity and quantum mechanics, via relativity of scale, by giving up the assumption of differentiability, or smoothness, and allowing for fractal roughness of spacetime across scales via fractal spacetime.

The mathematical intricacies of scale relativity and fractal spacetime, are largely beyond the philosophical scope of this book. While this prevents a detailed summary of Nottale's work, a general overview and several key points suggest the possibility of a rigorous mathematical undergirding to the fractal model of time presented here. Nottale's insights into temporal reversibility and fractal dimensionality in particular assist in developing this theoretical framework.

Nottale offers fractal spacetime as a method for establishing a relativity of scale. He defines the term fractal as: divergent with decreasing scale, building on Mandelbrot's definition, which highlights irregularity at all scales, and fractal dimensionality.[2] Divergent means to go to infinity. Therefore, scale divergence means that the more you zoom in, the more detail emerges—the infinite lengths within the coastline paradox. As discussed before, when measuring something fractal, the smaller your measuring stick, the longer the distance you will measure. This is scale divergence. A second key fractal property, related to scale divergence, is non-

---

[1] (Penrose 1994, 388–91)
[2] (Nottale 1993, 20, 33)

differentiability, meaning that a line is irregular or fragmented at all scales. It is never straight.

Scale divergence emerges as Nottale generalizes spacetime by removing the assumption of differentiability. Cosmology continually strives for greater generality in describing spacetime. A more general description encompasses more possibilities and makes fewer assumptions. For example, general relativity moved to curved spacetime geometry from less general, flat spacetime geometry. By challenging the assumption of flatness, general relativity describes how mass curves spacetime. By challenging the assumption of differentiability at small scales, Nottale opens a possible way to integrate quantum mechanics into cosmology.

Nottale points out that, like fractal spacetime, the paths of quantum particles are also continuous and non-differential. In fractal spacetime, the geodesics (straightest spacetime path between two points, as traversed by light) are also fractal–jagged and irregular. Thus, particles following fractal trajectories, e.g., Brownian motion, simply follows the underlying undulations of fractal spacetime geodesics.[1]

Nottale also notes that in quantum mechanics charge and energy diverge, or go to infinity, as scale decreases, thus providing an additional parallel between the quantum and the fractal. The divergence of energy with decreasing scale links to Heisenberg's Uncertainty Principle. In the Uncertainty Principle, the more accurately one variable (distance or time) is known, the less accurately its paired variable (momentum or energy) is known. This dynamic mirrors the relationship of scale divergence, in that the shrinking of one variable implies the divergence of the other. Nottale deduces that this divergence begins at the de Broglie scale, above which we find classical differentiable spacetime and below which we find quantum fractal spacetime characterized by temporal reversibility and fractal dimensionality.

Other physicists have also linked fractals to quantum realities. Physicist Tim Palmer postulated that the universe has an underlying, quantum, fractal invariant set which contains all the possible states of the universe.[2] Fractals are the invariant sets of chaotic systems. Invariant sets are sets describable by the least number of variables and parallel Nottale's notion of fractal spacetime as a more general form of spacetime by giving up differentiability.

---

[1] (Nottale 1993, 89-91)
[2] (Buchanan 2009)

Nottale's fractal spacetime recreates many pre-existing physical formulas through fractal re-formulation and offers a number of promising possibilities for testing. This book does not claim to ascertain the validity of Nottale's theory, however. The point of this book is that fractals are uniquely qualified to integrate many perspectives on time. Nottale's theory offers some mathematical formalisms with which to accomplish this.

Nottale focuses on developing fractal spacetime within the frameworks of mathematics and physics.[1] He leaves open for interpretation how this model might manifest in the human experience of time.

3.2.1 Scale Relativity

While fractal spacetime is a central tenet of Nottale's theory, he employs it as a method to achieve the larger theory of scale relativity. The notion of scale relativity extends our current notion of relativity. Relativity ensures that equations function equally well despite changes in location or velocity of observers. Nottale extends this so that equations will be equally applicable at any scale of application, regardless of the transformations of dilation or contraction.[2]

This formulation could be especially beneficial for the reconciliation of the quantum and relativistic, as scale divides them. In scale relativity, scale plays an analogous role to that of velocity in special relativity. Neither velocity nor scale can be defined in an absolute sense, but rather require a relational definition. Something is only fast or small in comparison to something else.

The nested, scale dependent nature of fractals makes them a natural choice for descriptions across scales. Additionally, fractals, as continuous but non-differentiable, achieve more generality than traditional geometries of continuous and differentiable spacetime. The next section explores the meaning of non-differentiability.

3.2.2 Non-differentiability

First, let us 'define differentiable. A differentiable function's graph has a definable slope, or rate of change, found by taking the derivative of the function. Where a graph changes direction, the slope becomes

---

[1] (Nottale 1993)
[2] (Nottale 1993)

undefined, the rate of change indefinite, and without derivative. Then the function is non-differentiable at the origin, point 0, 0.

The line in the right half of the figure has a slope (rise over run) of one and the line on the left half has a slope of negative one. At the point in the middle where the two lines meet, the rate of change is undefinable, and thus the function at this point is non-differentiable; there is no derivative.

If the vertical y-axis represents time and the horizontal x-axis represents space, then the first derivative, or slope of a line, describes the change in space with respect to time, or the velocity. The second derivative describes the change in velocity with respect to time, or the acceleration. When the line is straight, then the slope, or velocity, is constant and there is no acceleration, or change in speed, thus the second derivative is zero. If the slope is changing then the speed is changing, thus the object is decelerating or accelerating. This just provides one example of how derivatives show up in the real world.

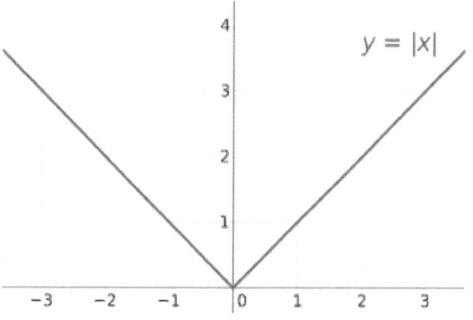

Figure 3.4 Non-differentiable at 0, 0.

The Weierstrass function, illustrated below, is an example of a function that is, like Nottale's fractal spacetime, both continuous and non-differentiable everywhere. Notably, it is also fractal. The closer one zooms in, the more obvious it becomes that every point on the graph is a point of inflection, where the graph changes direction, and thus it is nowhere differentiable. This non-differentiability is the roughness and unpredictability Nottale claims as the most general description of reality, not coarse grained into a smoothness that doesn't really exist.

A FRACTAL TOPOLOGY OF TIME

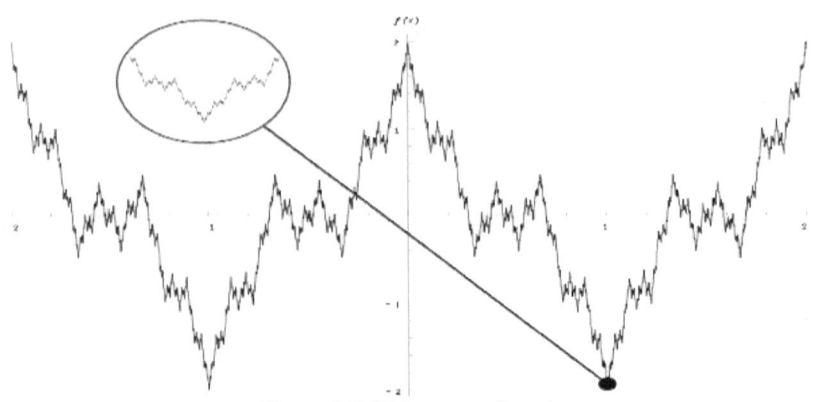

Figure 3.5 Weierstrass function
fractal, continuous everywhere, differentiable nowhere.

3.2.3 Fractals in Nature, Optimization

Figure 3.6 Rocky Mountains, Glacier National Park

Fractal patterning is found everywhere in nature, as in unfurling ferns, a river delta, or the jaggedness of a mountain horizon. Nottale takes this as evidence to argue that since fractal geometry is more general than either flat or curved spacetime geometry, "fractals would then be the structural manifestation of the fundamental non-differentiability of Nature."[1] Non-differentiable space-time would naturally manifest patterns of matter and energy patterns in fractal forms.

In the same way that the curved spacetime of Riemannian geometry includes flat spacetime of Euclidian geometry, fractal geometry includes them both. Fractal spacetime maintains the continuity of spacetime but becomes more general than differential geometries by giving up differentiability.2 The frequent occurrence of fractal structures in nature suggests that fractal spacetime might underlie this structural commonality, and perhaps even serve a common purpose of optimization, as Nottale

---

[1] (Nottale 1993, 39)
[2] (LeFevre 2009)

suggests. Nottale explores the benefits of fractals in nature and sees fractal behavior

> as an optimisation process. . . . Such a combination of divergence and boundedness may come from a process of optimization under constraint, or more generally of optimization of several quantities, sometimes apparently contradictory. Assume for example that some process leads to maximizing the surface while minimizing the volume, then a solution which optimizes both constraints is a fractal of dimension larger than 2 but smaller that 3 (infinite surface and infinitesimal volume).[1]

For example, repeating a branching pattern at smaller and smaller scales can build a tree, from the trunk dividing into two main branches, down to each individual leaf branching off, and even the veins branching within the leaves. This maximizes the tree's surface area, enhancing its ability to collect sunlight.

Just as the edge of the Mandelbrot set continually dances between bounded and unbounded points, resisting being pinned down as one or the other, no matter the resolution, so does probing the boundaries of nature reveal a similar unfurling of complexity at the interface of duality. This patterning is so prevalent in nature that permaculture[2] dubs it the "edge effect," describing the increased complexity occurring at the interface of different ecosystems. For example, the edge of a pond will host a greater variety and quantity of life than either the center of the pond or the field that surrounds it. The interface of these two ecosystems encourages creative ways to utilize the resources of both.

Life on earth is the fractal manifestation of the edge effect between matter and light. Water elaborates this dynamic tension, rising toward the sun and falling back to the earth. Plants construct themselves using minerals from the earth and energy from the sun, thereby increasing the surface area, or fractal density, of the earth.

One can extend this analogy even further to see our experience as the edge effect between time and timelessness. Recall the timelessness of photon traveling at the speed of light. Matter, as distinct from light on the other hand, exists in time. Thus, any interaction between matter and energy

---

[1] (Nottale 1993, 40)
[2] Permaculture is a method of cultivating abundance by assisting nature's preexisting tendencies and relationships.

(photons) is an interaction between time and timelessness.

Our brains provide one such site where the interaction between time and timelessness, perhaps via quantum processes, becomes so subtle that it is difficult to tease apart. The more closely we try to probe the intertwining of light and matter, thought and body, wave and particle, superposition and collapse, timelessness and time, the further their intertwining recedes, unfurling mysterious complexity in their wake, much as the Mandelbrot set's boundary between the bounded and unbounded unfurls as we zoom in. As conscious beings, we continually push the edges of the universe's growth. By exploring and cultivating these edges, we will discover abundance.

3.3 Timelessness and Fractal Dimensionality

3.3.1 Scale Divergence and Uncertainty

With this basic understanding of fractals, we can now take a look at how Nottale addresses time within his description of fractal spacetime. In the previous chapter, we reviewed several cases for the backwards flow of time, including closed timelike loops and antiparticles. Nottale's model also incorporates temporal reversibility, at very small scales.

Nottale takes the implications of virtual particles traveling backward in time to another level. He looks more closely at the possibility that what appears as many particles may, in fact, be one particle bouncing back and forth in time. In the figure on the left, one might see a particle moving forward and

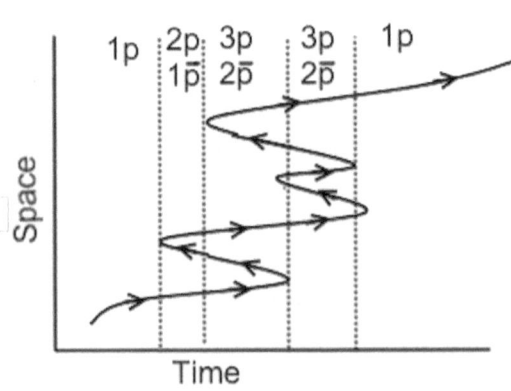

Figure 3.7. Particle time travel.

backwards in time. Alternately, imagine that you are looking at this scenario from a moment inside of time rather than the external perspective afforded by making it an image on this page. The dotted vertical lines delineate slices in time. From within one of the middle slices of time it appears that there is more than one particle, p, present in that moment. The "backwards moving" portions are antiparticles (p with a line over it) as discussed in section 2.2.2. Nottale assumes the one particle perspective, allowing it to return to the past for very short time periods.

> If these ... particles are actually manifestations of the same initial object, a correct calculation of the length it traveled and the proper time elapsed can only be done by adding the lengths and the times attributed to each individual particle."[1]

This assumption explains the energy divergence that occurs on quantum scales and fits the fractal characteristic of scale divergence.

We know that energy diverges in the quantum realm because of Heisenberg's time-energy uncertainty principle, as discussed in section 1.3.3. It describes that the more accurate, or smaller, the measure of time, t, is, the larger the uncertainty of the energy, E. Where Δ indicates change, or range of uncertainty, and $\hbar$ is Planck's constant.

$$\Delta E \Delta t \geq \hbar/2$$

The larger the energetic uncertainty the more virtual particle pairs are implicated as participating in the one particle's path to account for the large amount of energy. Then the proper time of the real particle must include the length of all the forwards and backwards paths of all the particles and antiparticles through time.

Nottale finds that the proper time traversed by this particle is independent of the timescale used for measurement, unless the unit of temporal measure is less that the de Broglie wavelength[2], at which point the proper time diverges as spacetime becomes fractal. The smaller the temporal measure, the larger the number of particles included, and the closer the particle's proper time gets to infinity. This divergence with decreasing scale is characteristically fractal and begins as the scale approaches quantum levels, specifically at the de Broglie time. The transition from the classical to the quantum is also a transition to scale dependence and fractal dimensionality. Nottale finds indicators in his calculations that suggest a "temporal

---

[1] *(Nottale 1993, 107)*
[2] A particle's de Broglie wavelength is Planck's constant divided by the particle's momentum.

transition from dimension 1 (non-fractal) to fractal dimension 2 around the de Broglie time = ℏ/E," where the proper time begins to diverge and becomes scale dependent.[1] Therefore, time is fractal at very small scales.

There are several known examples of fractal dimension of 2: the Mandelbrot set, Julia set, and Peano Curve, to name a few. These are space-filling curves because they cover the entire plane, thus earning their second dimension.

3.3.2 *Thickening of the Quantum Temporal Dimension*

Now what does it mean for time to shift from its classical single dimension to a fractal dimension of two, or visa versa? As the divisions of time become smaller, time's fractal nature emerges, creating an extra dimensional temporal thickness. Nottale labels this thickness as the axes t' and t" perpendicular to t. He does not address their physical meaning but acknowledges that they might actually be spatial dimensions. It seems that t, t', and t" would indicate three dimensions, but a fractal dimension of three would solidly fill the three dimensions. Three dimensions are required to illustrate a fractal dimension of two or higher. Since D=2 indicates that the fractal fills an entire plane solidly, the plane needs an additional dimension to expand into in order to gain perspective on its internal structure.

Building on the possibility that dimensions of temporal thickness are taken as spatial dimensions, t' = x and t" = y, I postulate a potential analogy to special relativity. As the divisions of time become smaller, proper time, τ, increases, and the trajectory's spatial length increases.

Proper time, τ, in special relativity, consists of both spatial and temporal dimensions in a formula akin to the Pythagorean theorem.

$$\Delta \tau^2 = \Delta x^2 + \Delta y^2 + \Delta z^2 - (c\Delta t)^2$$

In Nottale's description, as the intervals of time become smaller, they expand into a fractal dimension taking on a certain, possibly spatial, "thickness" or density. In a sense, at sub-de Broglie times, spacetime becomes more spatially and less temporally distributed. Similarly, special relativity indicates that increased speed in one frame of reference contracts the temporal and spatial units in a resting frame of reference, and vice versa.

---

[1] (Nottale 1993, 107)

In a sense, the spacetime density, of one frame with respect to the other, increases. This complements the increase in spacetime density facilitated by fractal spacetime.

In the above paragraphs, we considered Nottale's second temporal dimension as spatial for the sake of analogy with special relativity. Now let's consider Nottale's additional temporal dimension as a dimension of timelessness. The addition of a dimension of timelessness offers an inclusive view, as postulated in section 2.3.3, that might contain the temporal reversibility as discussed in the next section.

3.3.3 *Non-local Coherence*

Nottale describes the quantum realm as delocalized and coherent, and the classical realm as localized and incoherent.[1] I argue that the timeless relativistic realm, like Nottale's understanding of the quantum realm, is also coherent and delocalized, in contrast to the classical.

The classical world is local, meaning things have definite locations and abide by the limitations of those locations interacting with other places only at the speed of light or less. Delocalized refers to the absence of definite location. Points no longer separated by space communicate instantaneously. More accurately, delocalization is a disintegration of spatiality and separation entirely, such that there are no "positions" to distinguish between, and no distinguishable objects to locate. Our usual boundaries and identities blur and disappear.

The etymology of coherence is to "stick or hold together." The physics definition refers to a fixed relationship between waves, such that their phase difference does not change. For example, waves that are in phase with each other are coherent, like two playground swings swinging in time with one another. If one swinger decides to get off and slows their swing down, the two lose their synchrony or coherence. Nottale illustrates the coherence, or orderliness, of the quantum world as follows:

> Once admitted that the quantum fundamental 'object' is a probability amplitude, one notes that its equation of evolution is extremely simple, linear and deterministic. The quantum world is correlated, highly coherent, and far less subjected to chaos than is the classical.[2]

---
[1] (Nottale 1993, 162)
[2] (Nottale 1993, 161)

In other words, if we can accept the quantum nature of an object as a field of probability, rather than a definite, solid "thing," then all proceeds quite logically. This coherency, however, is not obvious from within object-oriented perception, which only considers the center of the probabilistic bell curve. While the object-oriented perspective explains some phenomena, the tail ends of the bell curve interject unpredictable behavior.

It seems the relativistic realm may hold a similar coherency to the quantum realm, as the timeless and spaceless nature of photons suggest an underlying unity. As discussed previously, photons, by virtue of traveling at the speed of light, exist in a unified timeless state. Existing outside of spacetime allows light to interact with any point of spacetime in a non-local way. Like other quantum particles, photons are non-local. In the quantum realm, position becomes blurrier. Whether you look at delocalization as points spreading out so that they overlap, or as all of space contracting into one point, either way, the distinction between separate points in space fades and the limitations of locality no longer hold. Considering the timeless, non-local nature of photons, the issue of action at a distance in the EPR paradox,[1] takes on new meaning. The photon's timeless nature, as a non-local hidden variable, seems to allow it to communicate with different points in spacetime instantaneously.

Secondly, one may view what appears as many particles at one moment in time, as one particle moving back and forth through time, utilizing the guise of the antiparticle when moving backwards in time. Following this argument to its extreme, one can imagine how all particles may be many manifestations of a single particle. For photons, specifically, the argument is stronger because of their timeless nature and reinforced by the fact that they are their own antiparticle. If a photon measures no spatial or temporal separation, then perhaps the photon appears as many photons only to a material, finite spacetime perspective. In this way, the relativistic realm is coherent by virtue of its unity. This seems to be a different type of coherence than that of the quantum realm, but there is definitely some overlap. This unified coherence, or unity of identity, is another way to think of non-locality. If there is only one particle existing in a state of timelessness, it is, of course, coherent with itself. It is only from our slower, perspective within spacetime that the photon assumes the guise of multiplicitous identities and locations.

---

[1] See section 1.3.4

Nottale characterizes the transition from classical to quantum via a fractal process called the Wiener process that describes random motion, including Brownian motion.[1] The squared Wiener process facilitates the transitions from quantum to classical, returning to the deterministic realm.

> We must give up the concepts of order and disorder as physically meaningful concepts and admit that both domains, classical and quantum, are organized, but differently. Then the D=2 Wiener process, W, becomes a reversible transformation (i.e., $W^2 = 1$) from one type of organization to the other. While delocalizing, W brings coherence, and while decohering, W relocalizes.[2]

Others before Nottale, including Bohm and Whitehead, have described this transition between the realms of the quantum and classical, the fractal and the differentiable. Nottale describes the transition from the fractal quantum to classical via the Wiener process. Quantum mechanics describes the transition from wave superposition to particle manifestation as the collapse of the wave function. David Bohm describes it as the unfolding of the implicate order into the explicate order.[3] Alfred North Whitehead describes the concrescence of an actual occasion from its prehensions.[4] This book draws attention to the emergence of unidirectional time from symmetrical timelessness. Fractal characteristics offer new perspectives to these descriptions of the quantum-classical transition.

Remarkably, Nottale's methodology also includes an account of the Wiener process's reversal, returning from the classical to the quantum. Understanding the transition between the hidden and the manifest brings us closer to an integral model of time and timelessness—how they fit together and how they interact. The next section explores another level of how timelessness and temporal reversibility relate to one another.

3.3.4   *Temporal Reversibility and Timelessness*

Nottale uses the fractal dimensions t' and t" to calculate the probabilities of particles moving forward (P_+) or backwards (P_-) in time. He finds that the smaller the temporal scale used ($\Delta t \rightarrow 0$), the more closely P_+

---

[1] (Nottale 1993, 143-153)
[2] (Nottale 1993, 162)
[3] (Bohm 1986)
[4] (Whitehead 1929)

= P_-, meaning that backwards or forwards time are equally probable at very small scales of time, with backwards time becoming less probable as scale increases and vanishing as the fractal 2-D time transitions to the classical 1-D time. For some particles, this might manifest as a balance of particles and antiparticles (interpreted as particles moving backwards in time as suggested in section 3.3.1).

"This result allows one to set the problem of the relations between microscopic reversibility and macroscopic irreversibility in a renewed manner. ...in the fractal model of time both of them naturally coexist."[1] This is exactly the model of integral time that this book advocates, acknowledging a variety of ways that time manifests and working to fit them together into a coherent whole.

Along the lines of the previous speculations of similarities between the quantum and relativistic, it is notable that the timelessness of light is only achievable by photon point particles. They exist in a state of timelessness. One could imagine that an absence of temporal participation might imply the limit of the temporal scale, $\Delta t=0$. If the probability of forwards and backwards time approaches equality with smaller and smaller time scales, perhaps their equal probability is found with photons, which have no time scale. Photons are their own antiparticle because they appear to behave the same backwards and forwards in time.

The previous paragraphs have illustrated that timelessness exists in the realms of the very small and the very fast. The timelessness and temporal reversibility of the quantum realm occur as timescales decrease and the dimension of time becomes fractal, spreading into possibly spatial dimensions.

3.3.5 *Fractal Dimension of Time*

The fractal dimension of time is not always two, however. Nottale finds, from the Heisenberg uncertainty principle and the energy of a free particle, that, "time in nonrelativistic quantum mechanics may *behave in some situations as a fractal dust of fractal dimension ½*, whose topological dimension is then zero."[2] In the quantum non-relativistic scenario, time's fractal dimension is ½. Nottale suggests that this is an artifact of the spatial

---

[1] (Nottale 1993, 163)
[2] (Nottale 1993, 104)

transition to D=2 before the temporal transition to D=2. The fractal dimension of time is two in quantum relativistic scenarios.

Figure 3.8 Cantor set

What does it mean to have a dimension of less than one? One can construct fractals by adding or deleting segments. Fractal dust occurs when deleting segments yields an infinite number of points of an infinitesimal length, like the Cantor set, above, where the middle third of each line in removed in each successive iteration.

One might experience any of these versions of temporality in as much as our perceptive capacities arise from quantum and relativistic mechanisms in the brain. While the notion of 2-D fractal time aligns with continuity and space filling infinite depth, the vanishing nature of 1/2-D fractal time also seems to capture something of temporal experience, in its non-spatial, intangibility. The slipping away of infinitesimal length captures the feeling of exclusivity of the moment generated by our inability to hold onto it. The paradox of time's slipperiness and its simultaneous inescapability finds expression in the mathematical phenomena of fractal dust. The concept of an infinite number of points corresponds nicely to the notion of an infinite number of moments in the expanse of time. Furthermore, the topological dimension of zero offers a possible explanation as to why time does not show up with the same spatial extension as the other spatial dimensions. Nottale goes on to point out that the conservation of energy depends on the uniformity of time and, thus, that the quantization of time could relate to the quantization of energy.

3.4 Imaginary Time

Complex, or imaginary, numbers play an important role in describing time in relativity, quantum mechanics, and fractals. Two main qualities of imaginary numbers make them particularly useful: providing a cycling pattern and distinguishing a 4th dimension to work with. These two qualities emerge from the complex plane, where the vertical axis is imaginary.

An imaginary number is the square root of -1, represented by i. A complex number is a combination of imaginary and real numbers, like i+4. Assigning the vertical axis to imaginary numbers, creates the complex plane.

To begin to understand the relationship between imaginary numbers and cyclicity, notice how each multiplication by i rotates 90° around the complex plane. Since $\sqrt{-1} = i$, multiplying i * i = -1.
 -1 * i = - i
 -i * i = 1
 1 * i = i
 i * i = -1

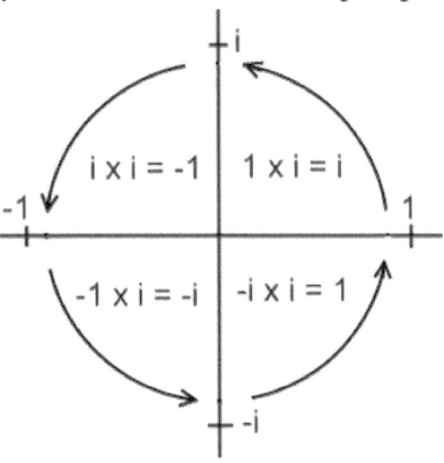

Figure 3.9 Wick Rotation

Each arrow in the image represents a multiplication by *i*. This is called Wick rotation. It takes four Wick rotations to return to the original angular position. Thus, successive multiples of *i* oscillate between the imaginary vertical axis and the real horizontal axis.

The cyclicity of Wick rotation comes into play in both quantum mechanics and fractals. In quantum mechanics, imaginary numbers facilitate the cyclic nature of the wave function and generate probabilistic predictions. Imaginary numbers' cyclic nature also contributes to the repetitions of fractal structure. Many fractals, like the Mandelbrot set, appear within the complex plane. Relativity utilizes the distinction between imaginary and real numbers to distinguish between time and space, while also bringing them into dialog.

Imaginary numbers represent time in quantum mechanics and relativity, (though could also apply to space in relativity). Most physicists assume that although imaginary numbers are useful, they have no physical meaning.[1] Noting both imaginary numbers' and time's cyclic nature (earth rotations, orbital, and clock cycles), the pairing of Wick rotation with time seems well-matched. In considering a fractal model of time, the role imaginary numbers play in the reiterative nature of some fractals may also provide some clues.

---

[1] (Jaroszkiewicz 2003, 163)

3.4.1 Quantum Mechanics

In quantum mechanics, when the Wick rotation makes time imaginary, the Schrodinger equation takes on the form of a real time diffusion equation, which describes how density disperses. In the imaginary formulation of the Schrödinger equation the square of the amplitude yields a classical probability for the location of the particle.[1] In other words it is the shift to imaginary time and back that gives quantum mechanics its predictive power.

Imaginary time also plays an important role in quantum field theory which relies on contour or path integrals occurring in the complex plane. The contour integral representation of the Feynman propagator occurs in the complex plane and, "shows that positive energy waves propagate forwards in time and negative energy waves propagate backwards in time."[2] Feynman interpreted this as anti-particles, like positrons, moving backwards in time while their particle counterparts, electrons, move forward in time. Here again, as with the block universe, the complex plane is offering us a view from outside of time.

The path integral formulation of quantum mechanics replaces the idea of a single path for a quantum particle, with the sum of all a particle's possible paths. Most of the particle paths cancel out leaving mainly non-differentiable paths close to the classical path. These then yield the complex amplitude, which is squared to yield the probability of the particle's path. This formulation is symmetrical in time and space.

In imaginary time the path integral becomes a Wiener integral, just like the Schrodinger equation becomes a diffusion equation. The Wiener integral is a common way to solve diffusion equations. Diffusion equations were first used to describe Brownian motion, the random motion of particles suspended in a liquid or gas. In the quantum case, diffusion seems to refer to the probability density of a particle's possible paths. Nottale derives the Schrodinger equation from classical mechanics via non-differentiable quantum paths and the implementation of the Wiener process.

> Start from quantum laws, add Brownian motion, and you get classical laws.... the reverse is also true: start from classical, add Brownian motion and you get Schrodinger's equation. This is an

---

[1] (Jaroszkiewicz 2003, 157)
[2] (Jaroszkiewicz 2003, 158) Or vice versa as formulated in the Dyson propagator. (Jaroszkiewicz 2016, 229)

extremely mysterious and paradoxical result. A Markov or Wiener process is usually considered as a method for describing disorder, and, in a naive way, one may expect that applying disorder twice would yield even more disorder. On the contrary, the combination of the results of stochastic quantum mechanics and the environment-induced decoherence seems to imply rather that the Wiener process transformation is "reversible"...[1]

Recall the duality of the quantum realm as delocalized and coherent, and the classical realm as localized and uncoherent, as discussed previously. Nottale goes on to sum up saying, "the classical and quantum domains are two 'orthogonal' types of organization which transform 'reversibly' one into the other by way of a Wiener process..."[2]

Jaroszkiewicz works with classical to quantum transition emphasizing the role of imaginary time in that transition more than Nottale's emphasis on the Wiener process. Jaroszkiewicz notes that, "The solution to the Schrodinger equation in imaginary time obtained via a Wiener integral gives the Feynman-Kac formula," which solves partial differential equations by simulating stochastic processes, and then, "gives Euclidean (imaginary time) Green's functions...for the harmonic oscillator."[3]

The use of the Wiener integral in the path integral formulation leads to the equation for a harmonic oscillator, that is the basis for quantum field theory, now formulated in imaginary time. In classical mechanics (real time) the harmonic oscillator describes oscillations between kinetic and potential energy, like the motion of a spring. In quantum mechanics the harmonic oscillator describes oscillations in probability.

When returned to real time via reverse Wick rotation, this yields physical predictions in real time. Jaroszkiewicz then states, "The great problem with this strategy is that it is not clear why going to imaginary time, doing Euclidean spacetime path integrals, and then returning to real time should be necessary in the first place."[4] He is not at all satisfied with this process which was initially employed to deal with poorly behaved path integrals. These singularities converge toward a definite value, but never actually reach one and therefore remain undefined.

Interestingly, Jaroszkiewicz mentions the similarity to

---

[1] (Nottale 1993, 161)
[2] (Nottale 1993, 162)
[3] (Jaroszkiewicz 2003, 160)
[4] (Jaroszkiewicz 2003, 160)

renormalization, the mathematical tricks employed to remove singularities. This provides another point of contact with Nottale's formulations. Nottale, however, defends undefined infinities as natural manifestations of fractal space-time and identifies how the effectiveness of renormalization illustrates this. Nottale identifies the role of the renormalization group in Wilson's many scales of length approach in developing scale dependent physical laws. He finds that fractal measures like length, area, and volume are solutions of renormalization group-like equations.[1] Nottale also demonstrates, "that the renormalization group equations ... can be interpreted as the simplest lowest order differential equations describing the measure on fractal geometry."[2] But my favorite is his use of fractal space-time to defend the existence of infinities, in this way preserving a space for the infinite of time - timelessness.

> Infinite numbers arise naturally in fractals, and the occurrence of infinite quantities is one of the difficulties of current quantum mechanics. It is remarkable that the infinities which appeared in quantum electrodynamics precisely concern physical quantities like masses (i.e. self-energy) or charges, which are fundamental invariants built from space-time and quantum phase symmetries. One may wonder whether the need for renormalization comes from the lack of account of the irreducible space-time infinities which would be part of the nature of a fractal space-time.[3]

Like Zeno's paradox, these infinites exist by virtue of the dimension of scale. Likewise, in the Mandelbrot set, squaring a complex number dives deeper in scale because a squared fraction gets smaller ($½^2 = ¼$) and squaring a complex number rotates it around the quadrants of the complex plane. These infinite depths are where the infinities live. If spacetime's fractal nature includes infinities at every point, it is well-positioned to integrate the paradoxical truths of mystical experiences, like experiences of timelessness or one's life flashing before their eyes.

Imaginary time has a natural relationship to periodicity or harmonic oscillation via Wick rotation, as seen in particle/anti-particle temporal oscillations. The periodicity of imaginary time is also utilized by Hawking and Hartle to define a universe that oscillates between expansion and

---

[1] (Nottale 1993, 2-3)
[2] (Nottale 1993, 24)
[3] (Nottale 1993, 28)

contraction avoiding the initial singularity, thus having no boundary, beginning, or end. "...Time is imaginary and therefore periodic. This is the Hartle-Hawking 'no boundary proposal', giving a picture of the early universe where the classical singularity is avoided...."[1]

This provides a further exploration of the temporal reversibility as a subset of timelessness as discussed previously. Then perhaps we can imagine how going to imaginary time might describe the interaction of temporal objects with the future and with timelessness, as Penrose describes in his concept of insight. Imaginary time might provide an opening for subtle influences of reverse causality.

Jaroszkiewicz seems stuck at an impasse between the timelessness of imaginary time and our experience of temporal flow.

> If the view is taken that the only thing of physical significance is the irreversible acquisition of information, which is in fact the only thing laboratory physics can deal with, then physics should be described by asymmetric process time rather than the symmetric manifold time of a geometric, block universe. Imaginary time lacks any asymmetry or arrow of time, which is the real cause for concern.[2]

I suggest that this conflict between temporal flow and timelessness is less a cause for concern, and more of a call toward investigating how the two concepts interpenetrate one another, as Nottale illustrates. I suggest linear temporality as a subset of reversible temporality, and reversible temporality as a subset of timelessness. The symmetry breaking of timelessness into asymmetrical temporality occurs not only as an event in cosmological history when energy folds into matter, but also continuously at the boundary where the classical emerges from the quantum.

Nottale postulates that most of quantum mystery can be traced back to the question "where does the complex plane of quantum mechanics lie?"[3] He points out that, "One of the most mysterious features of this theory was the complex probability amplitude: being irreducible to classical laws, it led to the belief that it is impossible to find its origin in space-time..."[4] Nottale, however, links the complex probability amplitude to fractal space-time which does not directly report to the classical authorities. He argues that the

---

[1] (Jaroszkiewicz 2003, 163)
[2] (Jaroszkiewicz 2003, 164)
[3] (Nottale 1993, 14)
[4] (Nottale 1993, 152)

complex plane naturally emerges from non-differentiable space-time.

3.5 Vrobel's Fractal Time

3.5.1 Fractal Measures

Fractals provide promising intuitive parallels to our subjective experience of time, which I began draw out of Nottale's descriptions. Which subjective experiences align with which fractal variables, however, requires a bit more teasing out. Theorist Susie Vrobel, in her 1998 publication, "Fractal Time," offers steps in this direction. She offers a fractal model of time that includes a notion of timelessness and our experience of time moving at different rates.

She identifies 5 measures of fractal time, based on fractal measures:

levels of description or LODs = levels of scale
temporal depth = number of scales / levels of description
temporal length = number of pattern iterations at one level of scale
temporal density = fractal dimension
condensation velocity = scaling factor

A level of description, LOD, defines a level of scale, which changes based on the measuring scale. Recall the coastline paradox where the coastline length comes out longer, the shorter the measuring stick. In this example, the respective levels of description (LODs) might be defined by measuring units like one mile, one yard, and one inch. In the case of time, levels of description may take the form of hours, minutes, seconds, stanzas of music, words, or syllables.

Cumulative LODs add up to temporal depth, t_{depth}. Figure 3.10 shows a progression through four levels of description, or units of temporal depth. Temporal depth describes number of scales of detail covered. For example, a piece of music made up of only quarter notes would have only 1 level of temporal depth, whereas the temporal depth available to a piece of music composed of whole note, half notes, quarter notes, eight notes and sixteenth notes would have greater temporal depth covering 5 LODs.

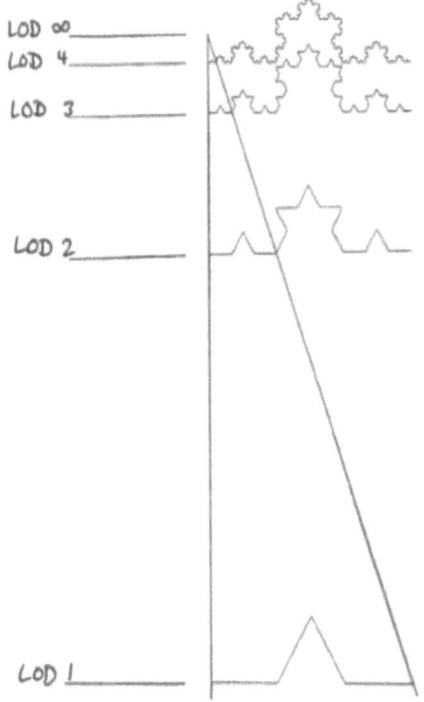

Figure 3.10 Levels of Descriptions (LODs)
(Vrobel, 1998)

Vrobel posits that the acts of recollecting, reflecting, and learning rearranges the past within the present, thereby increasing the LODs, and thus temporal depth. In this view, subjectivity generates temporal depth via memory. The more LODs, or memories, activated, the greater the depth of time, the richer the experience, and the faster time seems to go.

Vrobel uses the example of a couple attending a high school reunion. The alum experiences the evening as passing quickly because their richer memories mean more LODs. The spouse, on the other hand, with fewer associations to draw on, might have a relatively shallow experience, with only a few LOD's activated thus an evening that seems to drag.

One might expect Vrobel to align temporal depth with temporal density, since fractal depth is a depth of scale and thus contributes to density. Vrobel, however, defines temporal density as the fractal dimension, which depends on the number of copies per scale and the scaling factor, or in her terms the temporal length and condensation velocity, rather than the number of scales included.

Vrobel aligns temporal length, t_{length}, with successive events, like verses in a song or days in a week. She counts temporal length by counting the number of repeated patterns at the same level of scale. Condensation velocity, $v(c)$, is a fractal's scaling factor, s, for scale invariant structures.[1] The scaling factor is the number of times smaller or larger one scale is than another. For Koch curve, the scaling factor, and thus condensation velocity, is three, since each smaller pattern is 1/3 the size of the large pattern.

Vrobel equates temporal density, $t_{density}$, with fractal dimension,

---

[1] (Vrobel, 1998, 37)

defined as: $log\ n/log\ s$, where $n =$ the number of copies of the original pattern for each scale (proportional to Vrobel's temporal length), and $s =$ the scaling factor (Vrobel's condensation velocity).

One might assume temporal density has to do with the perceived rate of temporal flow. But Vrobel associates temporal flow rate with the number of LOD's (temporal depth) rather than with temporal density. The scaling factor, s, as used to determine fractal dimension and thus fractal density, does not count LOD's, but rather quantifies the density of their packing. Vrobel identifies the number of patterns per scale, n or a multiple of n, as successive events contributing to temporal length. Thus, her temporal density seems to be a measure of temporal length per temporal depth. Fractal dimension on the other hand is defined by number of copies per scaling factor, suggesting the temporal density as a measure of number of copies per rate of compression. At the end of this chapter, I elaborate on this alternative interpretation, where each compression is a step in time and the number of copies are the number of memories of a moment.

3.5.2   Timelessness, Insight, and Fractal Prime

Vrobel goes on to make connections with Penrose's concept of insight. Recall that Penrose suggests insight allows one to grasp an idea in an instant through contact with the timeless realm of Plato's ideas. Vrobel associates timelessness with the smallest iteration of a pattern, or the fractal's "prime", due to its indivisibility.

Often a fractal will have a lower and/or upper limit to scale invariance. Vrobel[1] identifies the fractal prime as the smallest nested structure from which further divisions do not occur. If a specific instance of the Koch curve did not divide below LOD 4, the iteration of the pattern on that level would be the prime. The prime is significant in a model of fractal time because the inability to generate further temporal depth denotes non-temporality, in the sense of lack of duration. Vrobel suggests that the timelessness of the prime facilitates Penrose's concept of insight occurring via contact with timelessness.

The non-temporal prime relates to all the self-similar structures in which it nests by virtue of their shared self-similar pattern. Storing information in compressed form, as in memories, facilitates the relationship between the larger external structures and the smaller, condensed, internal

---

[1] (Vrobel 1998)

structures. Vrobel refers to the larger-to-smaller relation as condensation, as occurs during learning, when information is compressed into memory, and when recalling previously compressed memories. Condensation is the subjective generation of nested LODs.

Vrobel aligns insight, as a relationship of timelessness to time, with the relationship of the prime to the larger structures. Insight zooms out rather than in, bringing an idea into being rather than making a memory. Instead of building time through the condensation of memory, insight recognizes the patterning across LODs. I suggest that, without the ability to compress information, we would not be able to grasp a pattern of information distributed in time, as a gestalt in one instant or "aha!" moment. Insight, or the ability to grasp a temporally distributed pattern in one instant, also allows us to identify one pattern with patterns on other LODs, thus bringing our insights garnered in one scenario to bear in novel scenarios.

In summary, think of condensation as--zooming in, learning, remembering--condensing information into memory to carry it into the future. Think of insight as zooming out, as the intuition that allows us to see the whole pattern in one glance, and relate patterns to one another, grasping the big picture. The folding and unfolding between the implicate and explicate orders sew the two together, building the structure of time.

As indivisible and timeless, the Vrobel's prime has a special relationship with the indivisible, timeless photon. Vrobel states, "the structure of the prime may be defined as a constant, in order to put all LODs into a relation to each other."[1] If the photon is the prime of a fractal, then what would the whole fractal look like? If we take the photon's pattern as wave / particle duality, we do find that this pattern repeats across scales via the Schrödinger equation.

Penrose[2] suggested that the phenomena of contacting timelessness via insight required a non-computational process and postulated the self-collapse of the wave function as one such process. Nottale mirrors some of Penrose's identification of the non-computability of quantum processes. Like Penrose, Nottale recognizes that Gödel's theorem, "has definitively demonstrated the logical difference between truth and demonstrability." Nottale also recognizes that, "some of the paradoxes of quantum mechanics reveal the first intrusion of this limitation into mathematical physics."[3]

---

[1] (Vrobel 1998, 48)
[2] (Penrose 1989)
[3] (Nottale 1993, 129)

With quantum mechanics we find non-demonstrable truths, such as wave-particle duality. Particles, by virtue of their wave nature carry non-local and thus non-temporal information. Penrose also associates the quantum realm with the timeless realm. Nottale also suggests that time transitions from one-dimensional to a two-dimensional fractal below the de Broglie length. I suggest that the second dimension that time's fractal curve fills in is the dimension of timelessness, which may also be considered spatial. One might object that the wave function evolves in time. But like photons' timeless nature still manifests in finite spacetime, so might the quantum realm participate in both.

One of the main differences between Vrobel and Nottale, as pertains to this discussion, is the fractal limit. Vrobel's indivisible, timeless prime is a lower limit, while Nottale's fractal spacetime has no lower limit, but is bound by an upper limit, which restricts his notion of fractal spacetime to the sub-de Broglie realms. Both however encounter timelessness at the smallest scales--Nottale at sub-de Broglie, Vrobel at the smallest version of the pattern.

Vrobel's theoretical prime however, may not necessarily have a physical correlate. The conceptual correlate of Vrobel's prime is the abstract idea that unites all of its individual instances--the "treeness" carried by every individual tree. Building on Vrobel's model, I also suggest characterizing the first experience of something as a prime--a baseline pattern that then encompasses every successive experience of that phenomena—in the way an experience of rejection can trigger one's primordial, childhood experience of rejection. As abstraction or primordial experience, the prime could be thought of as smaller or larger than individual instances—smaller in the sense that it is further away, or larger in its omnipresence.

Timelessness is both smaller and larger than us. That which is smallest, also surrounds us by virtue of its multiplicity. In this sense, we are encompassed by and built of tiny timelessnesses. Time both deepens into timelessness and sprouts out of timelessness. Each moment is both timeless, as well as an infinitesimal part of timelessness, both indivisible and continually divided.

Vrobel assigns timelessness to the prime, because of its indivisibility, I suggest that the pattern itself, no matter its divisibility, is timeless, in that every moment is timeless. I suggest that if the process of condensation that builds time, repeating patterns as memory, at any scale, is timeless, rather than just the smallest pattern. The process of time is made up of the repetition and compression of timeless moments, like condensation or deepening fractal patterns. In this way timelessness is not just a limit of time, but rather the container, the content, and the limits, if there are limits. This makes the conflict between Vrobel's lower limiting scale and Nottale's upper limiting scale less of an issue. Timelessness is the fractal's unchanging nature, illustrated in repeated patterns, across changes in scale.

Learning, in a sense, is a search for primes, for patterns of information that apply in different scenarios. Consciousness combs the furthest reaches of reality's details hoping to discover primes to organize the multiplicity. The mind condenses external data into memory. Stepping outside of that data, as if into another dimension, makes patterns and relationships in the data visible. The ability to step outside of our data to identify these patterns is the ability to step outside of time. Every time we compare two timeless moments, we bring them into relationship, building time by stitching together timeless moments. This is how we relate time to timelessness, by bringing the past into the present. This is how we identify patterns, by using the timelessness of the present moment to step outside the repeated patterns we have built into time, in order to view the timeless pattern.

We also relate time to timelessness by bringing the future into the present when we consider what might happen, making predictions and decisions. This is where the past and the future collide in the timeless moment that creates time through its self-condensation.

There are two horizons. The external horizon of the moment contains our present experience with all the possibilities of what to remember. The internal horizon of knowledge contains our collection of condensed memories. The horizon of the moment constitutes a boundary of information loss. We cannot retain complete information from the present moment. As we condense moments internally, they expand the horizon of our knowledge and increase its density. As memories recede in scale, their resolution decreases as well. Thus, age broadens the horizon of knowledge, while lowering the resolution, or reliability, of your memory. A wider horizon of knowledge, however, also increases your ability for insight.

3.5.3 *Subjective and Objective Time*

Vrobel has made excellent headway in beginning to quantify some very subtle and slippery concepts. I will now cover several topics that Vrobel does not address, in hopes of bringing some additional clarity to the theory. First, her measure of temporal length is a measure of *subjective* temporal length. She claims that subjectivity has an effect on the structure of time, but that subjectivity does not generate time itself. Thus, she finds objective time as more primary than subjective time, rather than believing that time is an artifact of consciousness along the lines of Kant. Vrobel's theory describes the effects of subjective time without establishing a place for objective time within the same system.

Objective time is consensus time, independent of a single perceiving subject. This convention allows people with very different senses of time to synchronize their activities according to regular external intervals, such as the rising and setting of the sun, or the linear ticking of a standard clock. Subjective time, as we know from experience, does not always conform to this linearity but instead often seems to go comparatively slower or faster. Objective time is synonymous with philosopher John McTaggart's B Series of earlier-later relations unrelated to a perceiving subject. Subjective time is more closely associated with his A Series, defined by past and future in relation to the continually changing present of a perceiving subject.

Vrobel asserts the primacy of objective time over subjective time. I find this idea limited, however, by the fact that we cannot know objective time, except through the lens of our subjective experience of time. The *experience* of consciousness is primary to the *idea* of time. Thus, the *experience* of the A Series (subjective time) is primary to the *idea* of the B Series (objective time). Additionally, I do not find it helpful to attempt to assign primacy to one of two such intertwined realities, not least because the very idea of primacy is itself dependent on temporal relationships. Non-temporality contains and supersedes the causal order of temporality. This does not make it primary but transcends the notion of primacy by transcending the notion of time.

In fact, this is where the primacy rested for McTaggart, in a combination of the A series and a non-temporal version of the B Series that he called the C Series. By identifying time with non-temporality, McTaggart adds fodder to his negation of the reality time. I agree with the reality of timelessness but disagree that it negates the existence of time. I believe that the question is less of time's existence or illusion, and more of defining its relationships appropriately. McTaggart proposes that time exists as an illusory subset of timelessness. I, however, do not find that timelessness negates the reality of time. Rather, timelessness and temporal reversibility provide the foundation for subjective time's unidirectional nature to emerge. How then, does the unidirectional flow of time interact with timelessness? How does consciousness fit into this interaction?

Vrobel does not offer mechanisms that might explain the generation of objective time, nor the relation of subjective speed of time to objective time. Nor does she attempt to utilize this model to describe time as external to subjective experience. I will attempt to address these issues in the following paragraphs.

3.5.4 *Temporal Density*

One of the paradoxes of time is that a particularly boring experience will seem to take forever while you are living through it, but in retrospect will seem to take up barely any time at all. In comparison, an exciting experience that seemed to pass very quickly, in retrospect seems to take longer than it actually did. Experiences condense into their essential components. Thus, a boring experience condenses more due to lack of meaningful components--vanishing in retrospect. Whereas, an exciting experience's abundance of meaningful components condenses less and expands in retrospect.

> Psychological studies show that our memories tend to retrospectively shrink empty minutes, hours and days while magnifying action-packed ones. In other words, time might seem to drag when we're bored, but our memories record just the opposite impression.[1]

I observe three measures of the same unit of time: clock time, experienced time, and retrospective time. The time a first day of school or work took according to the clock, differs from how long it felt during the experience, and from how long it seemed to take in retrospect.

Experienced time and retrospective time are inversely proportional by virtue of the experience's information density. Information processed per unit of clock time establishes the density of time. A segment of time packed full of information, passes quickly, though expands in retrospect in proportion to the amount of information processed. The more content included, the more highly divided the unit of time, the smaller the divisions,

---
[1] *(McCrone 1997, 52)*

the less time for each bit of content, resulting in an experience of a quickening pace of time. However, one recalls that time period as taking a long time due to the high information content.

On the other hand, largely empty segments of time feel as if they take a long time to pass but are recalled as being very short because of the lack of content. One might also see this as a measure of frequency, or beats per second, in that the smaller the divisions the denser the information. One may also take information's bits per second as another measure of temporal density.

Vrobel relates the rate of temporal flow to the number of scales, but defines temporal density as the fractal dimension, without relating it to temporal flow. I find fractal dimension as a better measure of temporal flow rate than temporal depth.

Recall that fractal dimension expresses how an object, when "crumpled up," takes up a portion of a higher dimension. A single line wraps back on itself to extend into a 2-D plane, like the Koch curve or the Mandelbrot set. A crumpled sheet of paper emerges from a 2-D sheet to an almost 3-D ball. Likewise, our bodies, and memories, fold back on themselves, slowing energetic flows and increasing our density into four dimensions of spacetime.

Time builds the density of bodies as memory folds the moment back onto itself through data compression. The equation for fractal dimension illustrates this folding back by dividing the number of copies, n, by the amount they are compressed, s. Fractal dimension = $\log n / \log s$, where n is the number of copies per level scale, and s is the scaling factor. Vrobel takes this as temporal length divided by condensation velocity. I propose an interpretation mode aligned with the definition of the variables in their fractal sense, with n as the number of memories of a moment and s as the rate of data compression.

3.5.4.1 Number of Copies as Number of Memories

The fractal variable n is the number of copies of the pattern per level of scale. Vrobel interprets n as proportional to temporal length. Since n is the number of copies of the same size, I prefer to consider the number of copies of a moment as the number of memories of that moment. The memory is a compressed version of the moment, but there will be many compressed memories of that original moment which might be considered to exist on a single level of scale. Many observers compress and store memories of each moment. The number of copies of a moment then equals or exceeds the number of observers of that moment. Of course, defining an observer and determining how many copies of a moment each observer might store presents a stumbling block for actual calculation, but could be simplified by definition as relative to specific scales.

The number of memories of a moment might include multiple memories stored by a single observer also. Individuals likely record more than one copy of a memory internally for the sake of redundancy. The more senses or ways of knowing a memory involves, the more "copies" of it get stored in different places in the brain. Each observer compresses each moment into one or more memories. The rate of data compression will differ for each observer.

Higher engagement in the moment corresponds to more copies and less compression per copy, which corresponds to higher fractal dimensionality. Curiosity and prior experience, among other variables, contribute to one's level of engagement. This might be quantified by brain wave frequencies, the basal ganglia loop's cycling rate, or dopamine quantities, as discussed in the following chapter. Higher rates of any of these correspond to increased divisions of time and higher rates of information processing, attention, or engagement.[1] These also relate to temporal perception. Lower engagement, or boredom, on the other hand, corresponds to fewer copies, greater compression per copy, and thus a lower fractal dimension. The higher the fractal dimension, the greater the density of time and the faster time's flow rate in the moment.

3.5.4.2 Scaling Factor as Rate of Data Compression

The scaling factor, s, is how many times smaller the next level pattern is than the pattern it nests within. The scaling factor characterizes the tightness of the levels' nesting. Vrobel refers to the scaling factor as the condensation velocity, condensing information into memory via learning.

I suggest that we might also consider the scaling factor as a rate of data compression. The rate of data compression increases with decreased novelty—as with boredom and/or age. When someone has had a similar experience many times before, they are likely to assume sameness and gloss over differences, compressing the successive memories of similar events

---

[1] (Sacks 1999; McCrone 1997)

more. Less novelty allows the moment to compress into a smaller memory. Thus, as novelty decreases with age, the density of information decreases, the rate of compression increases, and time seems to pass more quickly. The rate of data compression determines how tightly levels of description nest together—the more novelty, the less compression, and greater density.

Vrobel identifies the number of Levels of Description with subjective speed of time but does not include the role of the scaling factor--how tightly the levels pack together—in the rate of temporal flow. The scaling factor becomes the unit of measure for the dimension of scale, the linear measure of a logarithmic traverse. The smaller the measure the more tightly packed the levels. The larger the measure the more loosely packed. The scaling factor or condensation velocity determines the density or spaciousness of time. A higher scaling factor designates larger steps between scales and thus greater spaciousness than a lower scaling factor might.

The tightness of the packing might relate to how many levels of scale one can access at once. For comparison, consider how our senses scale logarithmically, allowing us to access a broader range of scale in sound and brightness. What looks just noticeably brighter or louder, is measurably many times brighter or louder.

For our senses the difference between apparent and absolute magnitudes is the difference between linear and exponential scales. The base of the exponent shows the scaling factor and the exponent shows the number scales. Together they show the total number of patterns replicated across that number of scales. The number of patterns increase exponentially as the number of scales increases linearly.

Taking the logarithm of an exponential scale yields a linear scale, translating between the two. The fact that we perceive exponential jumps as linear allows us to access a wider range of data by making bigger jumps for higher measures, while smaller jumps for lower measures reveal more detail on finer scales. Similarly, I suspect the more tightly layers of time nest together the more we can access at once. If they nest more spaciously, we may access fewer at once.

Likewise, time is a linear measure of an exponential flow of information/energy. As one moment gets compressed and recorded in multiple memories, information density increases exponentially. As information increases exponentially via collective memory, time progresses linearly, defining a logarithmic relationship between the two. The progression of time is the linear tick of the exponent, measuring the subjective torrent of information replication.

Just as time's linear progress glosses over exponential information flows, likewise the linear count across levels of scales glosses over exponential jumps in detail. Taking time as a dimension of scale offers a new understanding of time as deepening. Deepening time offers context for understanding its simultaneous exponential and linear nature. Since our senses scale logarithmically, perceiving a linear scale instead of an exponential one, it makes sense that we might process time in a similar manner.

The number of levels of scale count linearly, as the number of copies increases exponentially, and the pattern size decreases exponentially. Taking the log of the number of copies, n, and of the scaling factor, s, flattens these exponential measures into linear increases. Their ratio then defines the fractal dimension ($D = \log n / \log s$) as replication over compression.

The dynamic between information replication and compression recalls Nottale's definition of a fractal as emerging from an underlying dynamic of "optimization under constraint."[1] The tension between producing and reducing redundancy preserves detail while conserving space. Redundancies, or identical instances, can be collapsed into a single copy. The mind does this by comparing memories of similar experiences, thereby decreasing the overall "space" needed to retain multiple similar memories. Yet, maintaining multiple copies of a single instance, like in DNA replication, produces a redundancy that increases the likelihood of at least one copy surviving intact. Multiple copies can also cause errors and thus introduce novelty. Preserving multiple copies also maintains differences, while collapsing redundancy makes for more efficient storage, but may gloss over or erase subtle differences.

The fidelity of the data compression also contributes to the rate of data compression or scaling factor. Scale invariant fractals replicate the pattern with complete fidelity despite dilation, like a "lossless" data compression algorithm. Self-similar fractals, on the other hand, maintain the gist of the pattern, though introduce variation across scales as well, like a "lossy" form of data compression. "Lossy" data compression can compress more tightly than lossless compression, by preserving less details. The data compression formed by memory is more likely self-similar, or lossy, than scale invariant, or lossless, because it retains certain information, but not all.

---

[1] Nottale's characterization of fractal

I extend Vrobel's proposal that memory creates temporal depth, to include wave function collapse as also creating temporal depth by compressing the wave function into particulate form. Further, I take temporal depth not just as a subjective measure, but as an objective count of time as well, like age. Each compression deepens time into timelessness. Steps in scale equal steps in time.

Vrobel counts linear, objective time through the number of patterns at the same level of scale, creating temporal length. In contrast, deepening time counts time as a dimension of scale, a linear count of exponential information flow.

Each moment is stored and nested within the next. Time deepens into the timeless moment. Dividing the unity of timelessness opens the dimension of time as the dimension of scale within timelessness. Each compression deepens the depth of scale. Similar to Zeno's paradox, this sets up a potential infinite deepening of time.

These compressions of memory pull time into us. The direction of time is in, in every direction. Memories deepen into our bodies. Bodies are built from memories – of ancestors' lives, of places loved, of plants eaten, the soil that nourished them, the rain that fell on them, *ad infinitum*. The past nests within the present as embodied memories, as remembering bodies.

4
Deep Transcendence

Having seen several ways fractal models can benefit our understanding of time, I will now explore how the scientific insights might interface with our subjective experience of time. Building on the fractal description of how time emerges from timelessness via the timeless self-similar pattern, I now explore how this occurs cosmologically and subjectively, through a theory of deepening time, focusing on the effects of the variables of repetition, attention, and intention.

Repetition, attention, and intention are our experiential versions of fractal properties discussed in the last chapter. The fractal measure, n, the number of self-similar structures per level of scale, models the repetition that builds time from division. I make a case for attention as the synchronization of time scales or frequencies. Intention, like Penrose's notion of insight as Vrobel's prime, has to do with drawing on a pattern that is unrestricted by scale, a relationship to the infinite potentiality of timelessness. A fractal model of time offers a launching pad for addressing our frenetic pace of life and rediscovering nourishment from restoring the modern human's relationship with timelessness in daily life.

4.1 Variables of Subjective Time

It is easy to point to the distinction between subjective and objective time. The clock on the wall measures objective time, while subjective time is our felt sense of the duration of an event. While we can easily and accurately assess how much clock time has passed for a particular event, rarely does our felt qualitative experience correspond to this quantitative measurement. Many variables contribute to this disjuncture.

Deep Transcendence

Many of the variables are common knowledge, reflected for instance in sayings such as, "Time flies when you're having fun." Others are more subtle, but all of them seem to hinge on one or more of three factors: repetition, attention, and intention.

Repetition seems to have a role in building time, independent of conscious experience, through lunar or solar cycles, for example. Repetition also plays a key role in conscious experience and in attention. The number of times you do something tends to affect the amount of attention required for that task. Once you have biked around the block one hundred times the trip around the block seems to take hardly any time at all. However, the first time requires all of your focus and may seem like it takes forever. Thus, one of the things that affect your experience of the time it takes to complete a task is the number of times you have done that task before.

Another related factor is how much attention you are paying to the task. Attention then is the second order layer of complexity that navigates the linear temporality built by repetition. Repetition and attention mirror the fractal measures that determine the fractals dimension, n, the number of pattern copies, and s, the scaling factor.

When a moment condenses into memory, it does not just condense once, it condenses in every entity that remembers that moment. Thus, the pattern of the moment repeats in as many memories of it exist. Additionally, some perceiving subjects might store the memory in more than one place to prevent the degradation of details due to its compression. People with synesthesia,[1] for example, tend to have exceptional memories, by storing memories in multiple locations via different senses. So, when they need to recall something, they can find it easier because it has many "handles" to grab it with. Our DNA uses this strategy as well. Multiple codons (sets of three base pairs that code for amino acids) code for the same amino acid, so if one is damaged, backup copies exist. This is also why we repeat actions we are trying to learn; repetition eliminates errors, ensures accuracy, and maintains details. All these forms of repetition relate to the fractal measure, n, the number of pattern copies. Each entity remembers a slightly different version of the moment, to varying degrees of detail. The communal memory preserves details.

---

[1] Synesthesia is the phenomenon where people tend to experience multiple sensations in response to one sensorial stimulus. For example, they might see colors when hearing music, or different numbers might have different flavors associated with them.

Memory detail, in a fractal model of time, is represented by, s, the scaling factor. A more detailed memory will compress less than a less detailed memory. The less detailed memory will compress more, making it less accessible or harder to recall. Memory quality relates to the quality and direction of attention. What we attend to we remember. The greater the attention the lower the scaling factor. The two are inversely proportional. Thus, dividing by the scaling factor, as we do in the equation for fractal dimension, is the same as multiplying by the level of attention.

The fractal dimension is the log of n divided by the log of s. Restated in terms of fractal time, the temporal density is the log of the number of times the moment was condensed, divided by the log of how much it was compressed. Temporal density equals the number of memories times the level of attention (since attention is the inverse of the scaling factor). So, the more memories stored and the more closely you pay attention the greater your density of time. The log of s is actually equal to one-step in time. As compression, or attention, increases exponentially, time proceeds linearly. Therefore, fractal temporal density is also the number of copies per time.

As a measure of microstates per macrostate, n may play an analogous role to probability in entropy measures as well. Significantly, entropy aligns with our experience of forward temporal movement. Thus, the parallel of power law formatting for equations of entropy and fractal dimension suggest a possible alliance between

Intention is the third order layer that takes into account the big picture and directs attention in its navigations accordingly. Intention is the action of insight, zooming out of the fractal, drawing from the atemporal perspective of the timeless fractal pattern to identify and compare patterns, to check in with the larger communal memory, and to utilizing the subtle influences of non-locality and a-temporality.

Perhaps through a more careful exploration of the interaction between subjective and objective time, we might approach a better understanding of both consciousness and time. This section uses the variables of repetition, attention, and intention, in combination with a theory of deepening time, to understand the relationship between time and timelessness, and consciousness and cosmology.

4.2 Repetition

Building on Vrobel's fractal model of time, I postulated the progression of time as a process of fractal deepening within the timeless present moment, fractal depth as generated by wave function collapse, and each moment as made up of an interference pattern of previous moments, which contains a condensed version of those previous moments. In this section, I expand on the notion of the collapse of the wave function as a pattern repeated self-similarly within fractal time. Wave function collapse condenses the infinite potentiality of superposition into a manifest reality that further condenses into a memory or some physical record of its existence. The present moment both nests within the physical remnants of past moments and carries these past moments within it in condensed forms such as memory.

Repetition often makes things seem to go faster. Often an experience, such as a drive or hike, will seem to take much longer during one's first experience than it takes for subsequent repetitions. The first time you see the landscape, there is a great deal of external data to attend to and assimilate. By the time of the next hike or drive, you have condensed the experience into its meaningful aspects and have an organized expectation of what you will experience and therefore you do not pay as close attention as you did the first time. Your attention can dissociate from the landscape and become more absorbed in your own thoughts or conversations with fellow travelers. If you are not attending to a process, it will seem to go very quickly. Have you ever driven a familiar route lost in thought only to realize that upon arriving you barely noticed anything about the drive? Thus, it seemed to take practically no time at all. Here repetition, by redirecting our attention as a function of repetition, can speed up one's subjective experience of time.

When learning new skills, the more you do something the less attention it requires, the less time it takes, the more things you can do at the same time. Since life seems to build these skills one on top of another and the ethic of our culture seems to be that more is better, we continually cram more "things" into less time increasing the "time density." An increase in time density corresponds to an increase in the perceived speed of time. Thus, as individuals age, and as humanity gets older, time seems to go faster, on two accounts.

One can understand how time speeds up with age as a function of proportions. For a five-year-old one year is one fifth of their life, whereas to someone who is one hundred, one year is a mere one hundredth of that person's life. By virtue of comparison, a year naturally seems shorter than a year did when at age five. One hundred divisions make smaller pieces than five divisions would.

These divisions, in one sense, exist end to end as we usually think of time. In another sense, they embed within one another. Each year is an improvisation on the themes set forth in previous years. Since each subsequent year can reference any preceding year, the preceding years seem somehow larger, and more influential through laying down patterns continually referenced as the years carry on. This illustrates a fractal deepening of time. I refer to this perspective as deepening time in that we deepen into our lifetimes just as we deepen into our bodies, places, and other people. Our well-trod paths deepen into the earth. Our frequently used neural pathways correspond to deeply entrenched patterns of behavior.

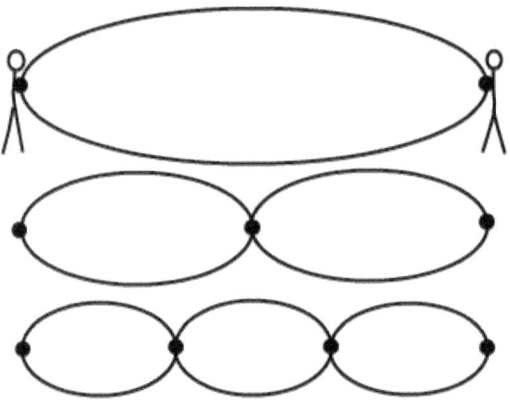

Figure 4.1 Harmonics: First, Second, and Third

To envision how years, or any division of time, can exist both end-to-end and within one another, imagine a vibrating string that might look similar to a jump rope spun by two people. The topmost image in the above figure shows an example of the first harmonic of a standing wave. It is half a wavelength. Increasing the frequency of the wave decreases the wavelength adding another nodal point in the middle, creating the second harmonic. The second harmonic includes an entire wavelength, both the crest and the trough. Increasing the frequency further creates a third harmonic, a fourth-nodal point, and includes one and a half wavelengths.

Think of the standing wave as the time between your birth at the left-hand node and your present at the right-hand node. You can imagine this for any point in your life such that the wave encompasses your entire lifetime from birth to death, or perhaps only the first year of your life. If you imagine the first harmonic, the top picture, as the first year of life, then once you have lived for two years you enter your second harmonic, the middle picture, which includes two smaller copies of the first pattern. Each year creates another division and a new harmonic for the wave. With each iteration, the first year's pattern is further compressed *and* replicated. Thus, the first year is both contained within the second year, represented by the first half of the wavelength, and maintains a sort of larger omnipresence by virtue of the nesting relationship of the first harmonic to the second. To imagine how the current moment is contained within the past moments, overlay the three harmonics of the figure above. The successive harmonics fit within the previous harmonics.

Note that this is reminiscent of the fractal property of self-similarity. Of course, the harmonics do not have to demarcate years per se but could denote any chosen measure of time. The division of scale creates what we perceive as the linear progression of time. A lifetime is one eternal moment divided into smaller moments. These smaller moments connect to the larger moment of a lifetime by their common pattern and nested relationship. Alternatively, in Vrobel's terminology the smallest, indivisible pattern, or prime, is timeless and omnipresent at every level of temporal scale by virtue of its shared pattern.

Another way to envision how the present is contained within the past is by taking a cosmological perspective. Because it takes time for light to travel to us from the far reaches of the universe, when we look out into space we are actually looking back in time. If we look far enough in any direction, we see back to the origin of the universe itself. Thus, we somehow nest within the point-like singularity that began the universe. One might also perceive this as the past embedding within our present. This is also true. The nesting is mutual, the past, present, and future mutually enfold one another. Inside and outside becomes a matter of perspective. In the same way, we can map the external sphere of the stars on a globe as if we were looking at the sphere from the outside instead of the inside. The singularity of the big bang exists both externally, at some finite point on the timeline of our past, and it surrounds us, and we continually deepen into it. The timeless quality of the moment allows it to both contain and be contained by time.

I began with familiar examples of how attention changes with repetition to introduce the section. I will return to the issue of attention specifically in the next section. For now, I will expand on how repetition might contribute not only to subjective time but to a sense of time as independent of consciousness as well.

4.2.1 Deepening Time

As stated at the end of the last chapter, Vrobel identifies temporal depth and LODs (levels of description/scale) as generated by subjectivity. I modify this claim in two respects. First, I equate temporal depth with linear or objective time, aligning the temporal dimension with the dimension of scale. Second, I suggest that subjectivity alone does not generate temporal depth, but rather navigates through layers of temporal flow nested within a preexisting timeless order. Subjectivity participates in temporal depth but is not the only method of generating additional nested layers. There are three places I will identify the process of timelessness deepening into time: in our individual lives, the evolution of consciousness, and cosmological evolution. First, I will try to convey a feel for the process itself.

I see LODs and temporal depth as a model of compression and repetition that relates a previous moment to the next. In a fractal, LODs emerge through the compression and replication of a pattern. In time, LODs emerge by the compression of information from a timeless present moment that is then nested within the following moment through a memory or the reverberation of its effects.

It is the explicit ability of consciousness to remember and to carry that memory into the future. Bohm anticipates a fractal structure of time in his discussion of memory:

> In each moment, there may be enfolded a memory of its past, in which, in turn, there may be enfolded a memory its past, and so on. Memory is thus in a kind of nested order of enfoldment (a bit like Chinese boxes).[1]

Repetition develops and facilitates memory. A child is not born with a notion of time. Someone must teach him or her this concept. In this sense children exist in closer proximity to a timeless state for a while, gradually learning to synchronize their internal clock with the temporal patterns of the external world, learning to associate and anticipate certain actions, and then to differentiate between repetitions of those actions.

The conceptualization of time relies on repetition and memory. The repetition of events eventually establishes a temporal structure of cyclicity

---

[1] (Bohm 1986, 181)

within the timeless moment. This cyclicity is not necessarily successive, but spatially returns to the same place. For example, one might consider each autumn as participating in an eternal autumn, like Plato's ideal form of autumn, which always corresponds to a particular segment of the earth's orbit. Only when we focus on the differences between one fall and the next do we separate one from the other, establishing them as two distinct entities that occur in succession. Thus, a notion of linear time flow emerges from cycles divided from one another through their own repetition.

Memory is not the only method for carrying the past into the future. The compression and replication processes occur in many ways. A tree's "memory" of the past years' rainfalls lives, physically, in its swollen cells generating thick tree ring to mark that year. Earth's "memory" of being struck by a meteorite lives, physically, in Crater Lake National Park. There are less obvious "memories" as well. We may not be able to deduce the reason a small pebble is here instead of there, but the pebble's location defines its "memory" of the foot that kicked it there.

I use quotation marks around the word "memory" because I do not intend to redefine the word to include more subtle expressions of its general concept, but to demonstrate how effects retain embedded information from their causes. This embedding of a previous reality within a current reality is the action of time.

It is likely that, evolutionarily, consciousness of temporality emerged in a similar manner as it does developmentally. Cultural historian Jean Gebser describes the transition of human consciousness from a timeless state, to a gradually differentiating dreamlike state, to cyclical time, linear time, and on into integral time (1986). Integral time incorporates the previous levels into a holistic picture, including timelessness, time, and the qualitative aspects of time. This book's fractal model of time aims to participate in the unfolding of this integral time.

The next section details how time also deepens within timelessness cosmologically.

4.2.2 Deepening Cosmological Time

There are other ways to understand deepening time as an interaction between time and timelessness. Since anything moving at the speed of light experiences no time, photons experience timelessness. Since it is impossible for anything with rest mass to accelerate to the speed of light, we know that matter, at macroscopic scales, does not participate in timelessness. Matter exists in time. Since photons exist in a state of timelessness, when one slows down, perhaps through absorption into an atom in a leaf, it "drops out" of timelessness, into time through its participation in matter.

Typically, when something exists outside of a particular dimension, it does so by participation in another dimension. If I stand in a line of people, I can only see the person ahead and behind me. If I step outside the line, I can then see everyone standing in the line. The line is one dimensional, stepping out perpendicular to the line I step into a second dimension.[1] Since photons exist outside of time and space, one can imagine that they might step out of these dimensions by participation in another dimension to which we are not privy. Just as a moving up high in the vertical dimension affords a broader view of the landscape below, I imagine the photon's atemporality affords it the broadest view of the temporal timeline. I think of a photon as outside of time, existing in an infinitely large moment that encompasses all of time. Then, when a photon interacts with matter, I think it actually divides this infinite moment, choosing a smaller nested moment to exist within thereby exiling parts of the previously infinite moment to the past and future. The photon deepens into time, like a higher octave in music.

With matter comes the differentiation of time and space into smaller, more defined portions of the infinite moment. Perhaps, in addition to flowing continually from the past to the present and on into the future, time also divides the eternal moment over and over again in a reiterative process. Like the perception of age, the more time passes, the smaller the divisions of time seem. The smaller the divisions, the more fit into the total lifetime, and the longer the total lifetime seems. Thus, successive division creates temporal flow. Division begets an appearance of linearity.

Since the big bang was the beginning of space and time, as we know it, any postulation of what came "before" the big bang would have to be outside of space and time.[2] In this sense, time emerged from timelessness. Furthermore, radiation dominated the early universe. The radiation energy density exceeded the matter energy density. Thus, the timelessness of photons also dominated the universe. As the universe expanded, the photon

---

[1] (Abbot 1884)
[2] Now the reader might notice that I have described a sequential process where I have also suggested the nonexistence of time. This seems paradoxical, and it is. It is a result of the edge effect of looking at and trying to describe a state of timelessness from within the process of time. One can only approximate the "other" from an experience of what it is not. We can describe timelessness to the best of our ability from within a temporal perspective but must keep in mind the limitations of this description.

density and individual photons energies decreased such that the total radiation energy density decreased faster than the matter energy density, until the matter energy density exceeded radiation energy density and the universe moved into our current period of matter dominance.[1] I suggest that with matter dominance came time dominance over timelessness. Matter facilitates division and thus the repetition that builds temporality.

In addition to matter's association with time, as distinct from photons' timelessness, it was largely the gravitational relationships of matter that taught us about time as our consciousness grew in its relationship with time. Specifically, the cycles of the mass and gravity of our solar system, give us our days, seasons, and years, from which we have come to know time.

Once the universe expanded and cooled enough to transition from radiance dominance to matter dominance, protons and electrons were able to combine allowing photons to decouple from matter and stream through the universe without continually interacting with other particles. This decoupling of radiation is still evident in the microwave background radiation, which fills the universe. Even though the big bang itself occurred 13.7 billion years ago, its remnant surrounds us, and we continue to deepen into it. When we look at the sky, we can still see the remnants of the matter-less energy from which we came. It does not exist on one side of us so that we are moving away from it, as linear time would suggest. It surrounds us, at the distance of the time it decoupled just less than 13.7 billion light years away. For, to look across space is to look back in time, even to the time of our birth as a universe.

Additionally, since every point in the universe was once the center of the universe, according to the Big Bang model, then the universe is its own center expanding away from itself. The universe is omnicentric. I suggest that the universe is not only omnicentric, but also omnigenetic (always beginning). In other words, just as time emerged from timelessness at the Big Bang, so is time born from timelessness again every time a particle/antiparticle pair is created, or a new life begins. Just like Vrobel's fractal prime, the initial act of creation, of time out of timelessness, cannot be further divided but rather propagates through division and multiplication. Time is continually beginning within itself.

Whenever energy folds itself into matter, as sunlight materializes into a leaf via photosynthesis, timelessness becomes time. Every time a wave function collapses into a particle, a moment differentiates from the

---

[1] (Bothun 2000)

timelessness of superposition. Time is continually reborn at every point it condenses within timelessness. As matter exists within a bath of energy, time exists immersed in timelessness. Timelessness also exists within time, in the infinity of each moment. The beginning of the universe is timelessness, not one point in time, but every point in time, continually beginning. Time is the continual deepening of the eternal timeless moment.

One can think of the universe as oscillating between matter and energy, and between time and a-temporality, as it continually manifests the present moment meeting between forward and backward time, between the past and future flowing into one another. We are the multiplicitous variations on the themes of these oscillations. Every moment that passes divides this one grand eternal moment repeatedly creating an exponential deepening, making the initial moment seem ever larger and larger as the universe's accelerating expansion.

4.2.3 *Abstraction and Quantum Collapse*

Whitehead[1] called this process of consolidating the infinite present into a smaller moment and shuffling it into the past *abstraction*. This philosophical description mirrors our subjective experience of time, and the physical processes of entropy and quantum collapse. David Bohm's work also echoes this theme. Building on the intertwining of these perspectives, I suggest that in the same way a fractal illustrates optimization within constraint, consciousness optimizes the amount of information carried into the future despite the constraints of finite spacetime.

In Whitehead's terminology *abstraction* is the process that gets an actual occasion, a moment of experience, from infinitely many prehensions to an objectified superject. Hence, abstraction is a form of our most concrete experience, the feeling of continual concrescence that is the basis for our perception of the passage of time. "The present fact has not the past fact with it in any sort of immediacy. The process of time veils the past below distinctive feeling."[2] Yet each abstraction embeds within it, via negative prehension, the entirety of the whole that contains it. While each new actual occasion has all of the past information embedded into it, its simplified manifest form seems to lose information through the process of time. Whitehead relied heavily on his own subjective experience and on the

---

[1] (Whitehead 1929)
[2] (Whitehead 1929, 517)

developments of quantum mechanics, so it is not surprising that this process, while intuitive to our subjective experience, also bears marked similarity with the process of the wave function collapse.

The theory of quantum consciousness links the processes of consciousness to the processes of quantum mechanics, specifically to the collapse of the wave function. In quantum mechanics, according to the standard interpretation, a system exists in a state of potentiality until it interacts with another system or, for our purposes, until we observe it. The interaction or observation corresponds with the mathematical event of wave function collapse. The collapse marks the system's move from a state of infinite potentiality, gradated by probability, to the concrete manifestation of one of the previously superposed (co-existing) potentialities. In the theory of quantum consciousness, this process maps onto our subjective experience of forming a thought, making a decision, or performing an action. These processes illustrate the pattern of a single manifestation emerging from a state of multiple possibilities. In the same manner, the present holds a certain potentiality that collapses into one concrete reality in the present. The present then condenses as memory and moves into the past, and consciousness moves into a new state of superposition. This process of superposition and collapse, superposition and collapse, repeats itself in a cyclic or reiterative process.

Hameroff specifically interprets wave function collapse as a mechanism that ratchets time forward. He notes that time emerges as the quantum becomes classical.[1] The quantum becomes classical through a measurement's felt physical effect. This is the collapse of the wave function. The difference here between Whitehead and Hameroff is that Whitehead's conception of forward motion in time through abstraction and concrescence is purely a unidirectional process. Hameroff, on the other hand, recognizes the temporal reversibility of the quantum realm, thus allowing for an acausal element in the process. The wave function collapse ratchets time forward classically from the timelessness of the quantum realm.

Our experience suggests that the limitations of consciousness prevent the perfect retention and accessibility of the infinite amounts of information in our past and future. Whitehead does not allow for reverse causality but does allow a place for timelessness within his cosmology. "In the temporal world it is the empirical fact that process entails loss: the past

---

[1] (Hameroff 2003)

is present under an abstraction. But there is no reason, of any metaphysical generality, why this should be the whole story."[1] For Whitehead, the rest of the story is the timelessness of the actual occasion, or moment of experience, which is internally non-temporal, but also the unit of temporality, and encompasses temporality through the actual occasion of God's eternal now.

Whitehead refers to the expanse of time, as described by the theory of the block universe, as God's concrescence. "God's concrescence extends over all time . . . is not subject to the passage of time . . . is not temporal...(and) does not perish."[2] Time, by contrast, seems to entail the perishing of information. I suggest, like Whitehead, that perfect information retention exists in timelessness that is present within time but veiled by time. I go beyond Whitehead in the suggestion that the future can influence the present through the medium of timelessness, appearing as indeterminate or acausal from behind the veils of our unidirectional perspective. We have limited accessibility to the infinite amount of information contained in timelessness by virtue of the limitations of out situatedness within time, and our related situated within consciousness.

The universal tendency toward increase of entropy mathematically describes the intuitive sense of loss of information through temporal process.[3] David Bohm expresses this necessary relationship between the loss of information and the progress of time.

> ...since loss of information is what is needed for establishing a new context, the increase of entropy with time may be said to allow later moments to constitute new contexts that are capable of having a certain relative independence from earlier moments.[4]

As Bohm states, loss of information separates moments, providing a limitation that facilitates the motion of time. Recall Nottale's characterization of fractal structure as a process of optimization under constraint, like a tree increasing its surface area with as many leaves as it can within its crown. If time is the constraint that prevents all information from occurring at once, there is something of an optimization process occurring for how much information can cross the threshold from one moment to the

---

[1] (Whitehead 1929, 517)
[2] (Hosinski 1993, 194-195)
[3] (Bohm 1986, 180)
[4] Ibid., 181.

next. Consciousness seems to facilitate the preservation of information, optimizing information transmission through temporal constraints.

Consciousness and time, by virtue of their mutual embeddedness, together facilitate and limit our perception of reality. Recognizing the limitations of our own consciousness, we can, from within our temporal perspective, step beyond that temporality to entertain a larger perspective of timelessness.

4.2.4 *The Influence of Quantum Timelessness*

Taking seriously the interpenetration of the realms of timelessness and temporal reversibility with unidirectional time yields a clearer understanding of the limitations of our temporally based perspective. When considering the timeless/temporally reversible nature of the quantum realm, all the information arriving at the present moment no longer comes solely from the past. Although predominantly inaccessible from within our time-bound perspectives, future information may influence our reality from the quantum level.

Quantum mechanics accurately predicts probabilities but cannot tell us which probability will emerge in any individual circumstance. The great debate in quantum mechanics is whether this is representative of the underlying reality of the situation or merely a limitation of our ability to know. I suggest that our temporal perspective limits our ability to know. By acknowledging a possible role for timelessness, and/or reverse causality, we might begin to hone our ability to identify its role and to utilize it in our decision-making processes.

Recall the EPR paradox, where changes to one particle affect the distant, entangled partner particle. By violating Bell's inequality, EPR experiments proved that the incompatible, or non-commuting, quantum observables, such as spin or polarization along different axis, are actually not determined until they are measured. If there are no hidden variables involved, then that leaves us to assume some violation of time, space, or the speed limit of light.

I am suggesting here that these violations do not appear so outlandish if one takes seriously the interpenetration of time and timelessness. If superposition and relativity involve a timeless component, then when particles entangle in a single state of superposition, their timelessness dissolves any temporal separation. They only re-enter participation with time upon measurement, whereby they also separate spatially.

From our usual perspective, the effects of interaction with timelessness might appear as faster than light information transmission, or perhaps even as information traveling backwards in time. In a sense, access to timelessness allows a quantum system access to the future from which it can bring back only a fraction of information.

If quantum systems are utilizing information from timelessness, or the future, to determine how they collapse, it is not surprising that this information is unavailable to our normal consciousness embedded in unidirectional time. Nor is it surprising that this information does not seem to exist in the past but seems to arrive when measured. The complete information in the present requires the input of the future, which only arrives in the present. Perhaps the mechanism of Aristotle's final cause (what a thing will be) lies in the timelessness of quantum realm.

Building on Penrose's link between quantum mechanics and the brain, Hameroff[1], like Vrobel, claims that consciousness creates time. I argue that, even without subjectivity, temporal depth and unidirectional time emerge with every wave function collapse. Thus, Planck time is Vrobel's fractal prime, the smallest nested pattern, indivisible and thus timeless. The ratio between our subjective time scales and external time scales then defines perceived temporal length. If time emerges with wave function collapse, as speculated by Hameroff, then it makes sense that time, as we know it, does not persist at sub-quantum levels. I recognize that consciousness participates in the emergence of time from timelessness, without necessarily being the sole driving force behind this process.

4.2.5   The Spectrum of Consciousness

I will take a moment here to define more specifically the spectrum of consciousness, as I see it. Because "consciousness" is such a broadly defined word, I find it more helpful to delineate various types of consciousness rather than split hairs over what is and is not conscious. I hope through this method to avoid alienating different ways of knowing.

When considering what consciousness actually is, I speculate that it hinges on our ability to project ourselves forwards and backwards in time, in other words, to plan for the future, remember the past, and make

---

[1] (Hameroff 2003)

connections between temporal moments. So, what is it that allows us to extend ourselves in the temporal dimension?

Consciousness, and temporal extension, seems to require separateness and a relationship with an "other." The roots of the word consciousness (and conscience), *con* meaning "with" or "together," and *scire* meaning "to know or to see," are put together to imply a relational knowing. This participation opposes the independent knowing implied by classical science, that shares the root word, *scire,* meaning "to know," but without sharing the notion of "with-ness."[1]

There are two main points I want to develop from this etymology. The notion of duality is important because it implies the linkage of two things germane to the development of my thesis: participation and temporality. "Withness" implies "two-ness" or a division. Therefore, in its most basic formulation, consciousness involves the participation of two separate things. Furthermore, division drives the deepening of temporality, and thus the progression of time.

I postulate three types of consciousness within its spectrum: experience, awareness, and self-awareness. These have temporal correlates. Experience exists in the timelessness of the present moment. Awareness utilizes memory to expand and divide the moment. Self-awareness further deepens the moment through the inclusion of future possibilities and choice. This spectrum also correlates to the variables of temporal experience used in this chapter. Experience is repetitive in its causal determinism. Awareness and attention closely intertwine, honing focus through memory. Self-awareness, through intention, empowers attention. When I refer to, or ask questions about, consciousness, one can infer the entire spectrum. However, in this book I am most concerned with self-awareness, as it is the least understood, the most encompassing, and therefore the most pertinent to this inquiry toward an integral understanding of time.

Before describing these types of consciousness in depth, I would like to dispel two common misconceptions about consciousness. First, just because we consider ourselves conscious beings, does not mean that we are 100% self-aware at all times. We tend to fluctuate though the spectrum of consciousness depending on the needs of particular situations and our own personal growth. Additionally, just because self-aware consciousness provides a powerful means of interaction with external reality, does not mean that it is superior to other modes of interaction. Consciousness, as I

---

[1] (Edinger 1984, 36)

use the term here, is a process of focus. Focus necessarily implies limitation and narrowing. This narrowing, or zooming in, also expands the details of the more delicate scales into which one penetrates. By recognizing the limitations of consciousness, paradoxically, we can begin to expand our self-awareness, even to include the largest conception of our self–the universe. Consciousness exists within the big picture, and, paradoxically, *as* the universal container of that big picture. In the same way, time and timelessness mutually contain one another.

The most basic type of consciousness is experience. All things have experience. That is, they sense their environment and they respond. So-called macroscopic inanimate objects, such as rocks and particles, sense and interact with the environment through physical forces, such as electromagnetism and gravitation. The interactions at this level are deterministic, repetitive, and participate entirely in the timelessness of the present moment. For inanimate objects, "memory" is physical and value neutral, contained in the object's shape, composition, and position. Experience does not contain value judgments.

All *living* things participate, to various extents, in the next type of consciousness, awareness. Living things have a vested interest in their survival and so value judgments color their experience. For example, plants seek out sunlight because it helps them grow. Plants pay attention to sunlight; they are aware of sunlight. Plant preferences persevere through memories embedded in their structural response. Evolution is the process of embedding memory of value judgments into the structure and reactions of a living thing. The embedded memory increases the LODs involved and expands the moment by carrying forward pieces of the past. In awareness, the focus of experience is narrowed by value judgments based on a vested interest in survival that is honed by an embedded memory that expands the moment. Memory and preference divide the timeless moment of experience, simultaneously deepening and expanding the timeless moment into temporality.

The third type of consciousness is self-awareness. Self-awareness includes not only the present and the past, but also develops the ability to take an external perspective on oneself and to imagine and choose possibilities for the future. Self-awareness parallels intention as a variable of temporal perception as outlined later in this chapter. It is empowering to become aware that you have an effect on your environment and that you can choose your actions based on your knowledge of these effects. This again adds more layers of thought, offering individuals simultaneous access to smaller divisions of time and to a larger perspective across time.

In the light of the indeterminacy of quantum mechanics, one might wonder where particles fall on the spectrum of consciousness. On one hand, I have just established a spectrum of consciousness that attributes only experience to inanimate objects, postulating that the interactions of inanimate objects are deterministic rather than value or choice driven. On the other hand, the indeterminate nature of the quantum realm seems to imply that a wave function might somehow "choose" into which state that it will collapse. However, choice is a property of the other end of the spectrum of consciousness that involves self-awareness. This is a bit of a conundrum, but one for which we are not ill equipped.

The self-aware end of the consciousness spectrum also includes physically embedded memory that divides and expands the moment. A particle, by virtue of the limits of its divisibility, cannot physically embed the vast amounts of memory typically associated with self-awareness. The notion of an expanded moment, however, is not foreign to particles. Notably, the photon's moment expands, not through the internal division of subjectivity that deepens time, but through its speed, which covers vast distances—thus encompassing more external, spatial division and dissolving temporal distinctions. The difference here is that the expanded moment of self-awareness in an internal moment deepened by the divisions of internal representation, and the timeless moment of a photon is an external moment, which might afford it an external perspective on spacetime. The timelessness of the photon encompasses all of eternity, whereas the timelessness of physical inanimate objects is a localized timelessness of the moment. The photon is thus a pivot point around which the spectrum of consciousness becomes a Mobius strip where, like experience, memory is still external, but like self-awareness, timelessness is internal.

4.3 Attention

As mentioned previously, repetition and attention intertwine closely. In the last section, I made a case for repetition as a basis of physical time. This section will discuss more specifically how the relationships between internal and external frequencies affect our experience of time. First, I will review the role of frequency in physics. Then I discuss internal time sense as a function of biological process involving temperature, the neurotransmitter dopamine, the brain's electrical frequencies, and the rate of quantum collapse. I then make the case for attention as a synchronization of one's internal clock with the time scale of the external object of attention.

Like wave function collapse, interactions and changes impose boundaries on moments. The present ceases to be the present when a wave function collapses--either through an external interaction, or of its own accord via Penrose's concept of objective reduction (OR). Thus, as the infinite potential of the wave function collapses into concrete manifestation, time emerges out of timelessness and is stored in physical objects. Consciousness is another level of this capacity of physical things to compress and store information. It "packages" a moment and stores it as memory.

Attention collapses a moment via internal or external change. Attention allows us to track and notice external change, while an internal shift in attention offers change by closing one moment and opening another, by compressing information to move it into the past and opening one's senses to the next moment. Attention tends to be directed from within, by intention, or commanded from without, by novelty or change. Therefore, change and attention together create the boundaries, or differences, which allow us to distinguish between "before" and "after." Attention provides the boundaries between successive repetitions, as well as links them together.

Attention navigates levels of description by synchronizing its internal LOD with the external LODs. The act of remembering shifts one's attention to a more deeply embedded LOD. Similarly, the act of focusing on one aspect of external reality relates one to deeper levels of description by virtue of the complexity of that object's embedded history. Every physical form in some way embeds its past information and carries it into the future.

4.3.1 Frequency

Any given segment of time contains a wide range of frequencies nested within it, from the slowest rotations of the galaxies to the fastest quantum collapses. Similarly, individual human frequency contains many layers of frequencies of different scales. Thus, it is a difficult task to characterize an internal or external frequency in one number.

Physics has taught us that all physical objects have an associated wave nature and thus an associated frequency. I will review a bit of wave mechanics terminology before proceeding.

Waves vary by wavelength, frequency, and amplitude. Amplitude has to do with the intensity of the wave and, in quantum mechanics, is squared to yield the probability of measurement at a particular location. The wavelength (λ) and the frequency (*f*) are inversely proportional.

$$\lambda = v/f$$

This means that as one increases the other decreases by the same factor, and vice versa. Frequency is measured in Hertz, or waves per second. So naturally the shorter the wavelength the more waves per second–the shorter the wavelength, the higher the frequency.

Even particles with mass have a wave nature, as defined by the deBroglie wavelength. Wavelength (λ) is inversely proportional to momentum (*p*).

$$\lambda \propto 1/p$$

Since momentum (*p*) is mass (*m*) times velocity (*v*),

$$p = mv$$

the larger the mass is, the shorter the wavelength and the higher the frequency.

$$\lambda \propto 1/mv \propto v/f$$

Then Schrödinger established the wave nature of any sized mass. If mass has a wave nature, it behaves differently than we typically think of mass as behaving, smearing it out in spacetime a little. Heisenberg's uncertainty principle limited the accuracy with which we can simultaneously know the position and momentum of a particle.

Time changes scale by changing frequency. Just as the wavelength decreases for an increase in frequency, the time period associated with the wavelength decreases. When the phase of short wavelengths of higher frequencies aligns with the longer wavelengths of lower frequencies, then the waves are

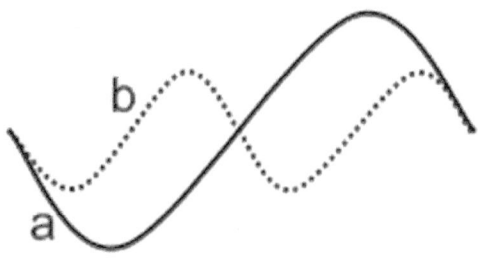

Figure 4.2 First Harmonic

harmonics of one another, similar to the way smaller fractal patterns nest within larger patterns.

Let us review how frequencies of different scales show up in nature. As speculated by Penrose, a mass of one kilogram could exist in superposition for only 10^{-37} seconds before it collapses of its own accord. That means that its wave function cycles between collapse and superposition 10^{37} times a second. Therefore, the wave function collapse of massive objects is probably one of the highest frequencies we are dealing with here.

The electromagnetic spectrum describes an infinite range of frequencies all propagated by photons traveling at the speed of light. At the highest frequencies, of 10^{20} Hz and above, there are short wavelength UV, X, and gamma rays. We can see the mid-range spectrum of visible light ranging from violet's 7.9×10^{14} Hz to red's 4×10^{14} Hz. Radio waves are the lowest frequencies and longest waves of the electromagnetic spectrum with frequencies around 1000 Hz.

Human hearing can detect sound frequencies[1] between 20,000 Hz (high pitched) and 20 Hz (low pitched). The electrical activity of the brain, measured by EEG, ranges from 100 Hz gamma waves associated with specific cognitive and motor functions to below 4 Hz for deep sleep delta waves. The human heart beats at about 1 Hz at rest and up to just over 2 Hz during intense activity. Then you arrive at the more familiar frequency ranges of seconds, minutes, hours, days, weeks, months, years, centuries, millennia, etc.

The notion of scale is applicable to the study of subjective time, since we have yet to quantify the various relationships between the qualitatively different time scales we experience, from seconds to lifetimes. Just as we understand position as relative to the position of the observer, it makes sense that we might measure time depending on the temporal scale of the observer (i.e., their age, or pace of life). This concept moves us away from a notion of absolute time and toward a mathematical model of subjective/relative time.

To flesh this model out I need to explore the variables that contribute to variability of subjective time. With that basic description of frequency, I can begin to look at what physical processes determine our internal and external frequencies and how they relate to one another. I look

---

[1] Sound frequencies differ from electromagnetic waves by what is waving. Sound waves are pressure waves which require a medium to propagate through as opposed to electromagnetic waves which are carried by photons and able to traverse a vacuum.

specifically at temperature and chemical reaction rates, brain frequencies, and the synchronization of entrainment.

4.3.2 Temperature and the Internal Clock

We are starting to understand how the mind processes time physically and chemically through the study of several mental disorders linked to alterations in one's perception of time. In this section, I explore neurological processes related to the perception of time. In the following section, I address how these might interface with the fractal model proposed in this book.

One thing that researchers look for in the brain is some sort of time-keeping mechanism. In early studies, Hudson Hoagland discovered that, "raising the brain's temperature can alter a person's sense of time by up to 20 per cent."[1] He found that time seems to go slower for people experiencing high brain temperatures via fever or external heating. When considering that temperature is actually a measure of the speed of particles and a measure of the rate of chemical reactions in living organisms, it makes sense that an increase in temperature would correspond to an increase in the rate of the brain's internal clock. If one's internal clock seems to run faster than external objective time, then external time will feel like it is running slower.

Hoagland was able to plot subjective perception of time against body temperature and found it to correlate with the Arrhenius equation that describes how chemical reaction rates vary with temperature. Independently, Heinz Von Foerester used the same equation to study memory and its rate of decay.[2] The equations of Hoagland and Von Foerester yielded the same measure for reaction rate, or energy per unit time, implying that "our time sense depends basically on rates of biological oxidation in the brain" and "consciousness depends upon awareness of duration."[3] In fact, Von Foerester's model of memory storage and decay was based on the maintenance and quantum-mechanical decay of certain protein molecule states,[4] in marked similarity to the Penrose-Hameroff model.

When considering Penrose's theory of consciousness as facilitated by quantum collapse of neural proteins, it makes sense that these collapses

---

[1] (McCrone 1997)
[2] (Hoagland 1981)
[3] (Hoagland 1981, 324)
[4] (Hoagland 1981, 312-329)

would occur with greater frequency in an environment where chemical reactions are happening faster. The faster neural proteins move into superposition, the faster they can reach the critical mass for self-collapse (see Figure 1.3).

Notably, some drugs, such as amphetamines, cocaine, LSD and MDMA, which alter one's perception of time are also pyrogenic or fever inducing indicating an increased rate of cerebral metabolism. This aligns with section 1.4.3's discussion of psychedelics increasing electron mobility and thus the rate of quantum collapse in the brains proteins. Metabolic rate is not the only factor in what may constitute an internal clock, however.

4.3.3 EEG and Scale Synchronization

Another variable that might come into play in the brain's internal clock is its electrical activity. Electroencephalography or EEG uses electrodes placed on the scalp to measure the electrical activity of the brain. It has shown how various mental states correspond to various subjective experiences.

The longest and slowest of the distinct electrical patterns produced by the brain are delta waves. Delta frequency, 0-4 Hz, or beats per second, emerge in babies and during deep sleep in adults. Theta waves (4-7 Hz) occur in young children and in drowsy or meditating adults. Alpha waves (8-12 Hz) indicate a person is awake and relaxed. Beta waves (12-25 Hz) correspond to alertness, activity and thinking. In addition, gamma waves (over 25 Hz) occur in correspondence to certain cognitive and motor functions. The 40 Hz signature in particular seems to play a role in unifying consciousness across the brain through synchronizing neural firing, as previously discussed in section 1.4.3.

The synchronization of neural firing that covers the brain during 40 Hz processing is a form of entrainment. Other forms of entrainment include when heart cells in adjoining Petri dishes gradually begin to beat in time with one another, or the menstrual cycles of women living together gradually align, or the brain waves of violinists playing the same piece match one another. I propose that just as entrainment within the brain can unify consciousness, that entrainment between the brain and some external frequency can affect both individual and communal perception of time. I argue that seeing attention as a form of this entrainment might offer one way to quantize the effects of variables on temporal perception. Like the violinists, paying close attention seems linked with brainwaves entrained to an external stimulus, whether the brainwaves of another person or the frequency of a material object. Experienced time depends on a ratio of internal to external time.

The rate of subjective progression of time depends on the ratio between the rate of one's internal processing compared to external processes. When one's mind is spinning and nothing seems to be happening externally, a moment seems to last forever. When one's internal and external frequencies align, no time seems to pass at all, similar to the trance like state that one might experience when dancing or drumming in time with a larger group. One's internal and external frequencies are always seeking to entrain with one another; the internal drives the external through intention and the external sets the pace for the internal by drawing attention.

One's internal frequencies drive external frequencies by cultivating their surroundings to match their pacing. If a person's mind is going a mile a minute, then they are likely to surround themselves with a multitude of technologies for instantaneous communication (e-mail, cell phone, news feeds) that will then reinforce their breakneck pacing by placing frequent demands on their attention. Additionally, a person accustomed to rapid-fire activity might try to impose this mode of business onto people who prefer to take life at a more leisurely pace creating conflict.

If we assist the natural process of entrainment by remaining attuned and attentive to people and patterns outside of ourselves, we can reap the benefits of an effortless relationship with time. If, however, we are overly self-focused and trying to use our internal frequency to drive unresponsive external frequencies, rather than synchronizing with them, we might be setting ourselves up for perpetual disappointment. This is a matter of intention.

Take for example the relationship between an adult and a baby. A baby's brain waves are predominantly delta waves, those that occur during deep sleep in adults. It is not surprising then that we tend to lack a developed sense of time in both infancy and during sleep. One might refer to these as states of timelessness. One example of attention as scale synchronization is how easy it is to slip into an experience of timelessness even when just playing with a baby. The baby is attuning itself to the adult world and the adult attunes itself to the child's world and rhythms.

Research at IHM (Institute of Heart Math) is showing that when a person is touching another or in close proximity to another, one person's cardiac rhythms can become coherent with the other person's cardiac rhythms as measured by electrocardiogram (ECG). In addition, that coherent ECG signal can be measured in the other person's brainwaves or electroencephalogram (EEG).[1]

By tuning into the baby's slow wave processing, it seems likely that one's own brain waves would drift in that direction. Since new mothers often do not get as much deep sleep due to the frequency of feeding, perhaps they get a dip into delta waves through sinking into baby presence during waking hours also. Additionally, as I discuss in a later section, dopamine (implicated in time perception) and prolactin (the hormone associated with lactation) inhibit one another, so while a mother is breastfeeding, her dopamine system is on standby, possibly decreasing accuracy of time estimation. Perhaps this combination of factors contributes to why a child's infancy seems to pass so quickly for mothers, because it involves more stepping outside of time than adult interactions. On the other hand, a person caught up in a high frequency lifestyle might have a hard time transitioning into a resonant state with such slow brain waves if they do not cultivate the intention to be present to the child.

In addition to the entrainment to delta waves, the complete engrossment in the task of child rearing would also lend itself to an experience of timelessness in the complete engagement of one's attention, largely demanded externally by the needs of the child. While a baby's own internal frequency might be comparatively slow, the frequency of their needs demands our attunement to them. When there is another person who keeps collapsing the wave function of your reverie with his or her need to be fed, changed, or played with, time both moves faster and maintains a certain momentum that keeps us closer to an adult pace of life. The combination of attending to both the slow frequencies of the child's experience with the higher frequency of parental responsibility demands functioning in multiple levels of description creating a density of time unlike tending to responsibilities for yourself alone.

The more content divides a moment, the greater the information density of that period of time. The number of divisions corresponds to frequency, measured in beats per second or Hertz (Hz). Higher brain wave frequencies correspond to greater activity, making time seem to pass faster than when our brains are idle and brainwaves are slower. Brainwaves do not

---

[1] (Pribram and Rozman 1997, para. 12)

occur in isolation. In fact, our perception of the speed of time depends largely on the relationship between our internal clock and the pace of life that surrounds us. Typically, the two seek to synchronize with one another. If you are living in a busy city, you will likely be busy as well. If you are living in a woodland cabin you are less likely to divide your day into as many micro-moments as you would with the higher levels of interaction that accompany city living.

4.3.4 Dopamine and the Internal Clock

Beyond the studies of brain temperature and EEGs, the neurotransmitter dopamine also plays a role in temporal perception. Warren Meck discovered that during temporal processing, a centralized structure called the basal ganglia activates.[1] This area regulates a circuit of neurons that loop around connecting many areas of the brain. Meck first postulated that the rate at which signals traverse these loops offer the brain a time keeping mechanism.

The substantia nigra, and the dopamine it produces, regulate the rate at which signals traverse this loop. The more dopamine present, the faster the pulses of the basal ganglia's neural circuit. Dopamine is a neurotransmitter that seems to play a key role in temporal perception, motivation, and facilitating continuity of thoughts and actions.

This would seem to suggest that more dopamine and a faster internal clock would be associated with a slower perception of external time, like the increases in brain temperature, and lower levels of dopamine and a slower rate of neural circuit frequency might correspond to a slower internal clock and a perceived acceleration of external time. The association, however, is not so clear-cut. The characteristic under or over activity of the dopamine system is linked with impairment in accurate time estimation in general, sometimes making it seem slower, sometimes faster, but there is no obvious correlation between perceived speed of time and the amount of dopamine present as one might suspect. First, I will explore some of the correlations between dopamine and experienced time.

Dopamine naturally decreases with age. This decrease likely accentuates the slowing of one's internal clock that accompanies a slowing metabolism, and the shortening of years once one has accumulated many

---

[1] (McCrone 1997)

with which to compare them. This combination of factors contributes to the apparent increase in the pace of external time that occurs with age.

Abnormal dopamine deficit plays a role in the jerky movements of Parkinson's and the abrupt attention shifts in ADHD folks. In the extreme, the absence of dopamine is the root of the extremely slowed or frozen states of patients described by Oliver Sacks in his book *Awakenings*.[1] By administering large doses of the dopamine-mimicking drug L-Dopa, Sacks restored these catatonic patients' ability to move and to participate in the world, and sometimes launched them into hyperactivity.

To these patients, time did not seem to pass any slower or faster. Sacks encountered a man in the hallway who held his hand frozen in midair, for what seemed like hours. The man reported that he was merely wiping his nose. Sure enough, with the help of time-lapse photography, Sacks was able to see that this man was actually performing such a task, just at an imperceptibly slow rate. Patients did not report feeling as if they had spent a long time in their frozen state, but just instead experienced a sort of timelessness, without perceiving the external motion buzzing around them.[2] "One woman described it as like living in a still pond forever reflecting itself. Her awareness of the world was bright, fixed and hard-edged like the picture in a stained-glass window, fantastically pure but empty of possibilities."[3] This is similar to how I would suspect that a photon, in its timeless state, would perceive the world, as frozen.

On the other end of the spectrum of activity levels, ADHD is also associated with decreased dopamine levels, decreased activity in the frontal lobe, basal ganglia, and cerebellum. These are all important for temporal perception. Their decrease manifests as impaired time estimation and a perception of time passing slower, hence ease of boredom, short attentions spans, and hyperactivity.

Ritalin and Adderall, both forms of amphetamines, boost dopamine activity and returns the brain functioning to normal levels, allowing children to become more engaged in what they are doing.[4] "Ritalin also works to suppress "background" firing of neurons not associated with task performance, allowing the brain to transmit a clearer signal."[5] It is easier to focus and pay attention when your brain is not being pulled down multiple

---

[1] (Sacks 1973)
[2] (Sacks 2004, 60-69)
[3] (McCrone 1997)
[4] (Coghlan 2009)
[5] (Brookhaven National Laboratory 2001, par. 8).

pathways. Moreover, if you can pay attention, you establish some traction between your internal clock and the external world by latching onto something outside of yourself, synchronizing with it, using it to help set your internal pace. If your brain is going in a million different directions you cannot attend to something outside of yourself long enough to synchronize with it, and you become isolated inside yourself making it difficult to learn.

On the other hand, schizophrenia, and amphetamine intoxication which causes symptoms similar to schizophrenia, cause errors in time estimation. Both seem to benefit from drugs that block dopamine, implying an overabundance of dopamine. Amphetamines can cause Parkinson's-like symptoms in long time users. A healthy person can distinguish the order of events between things that happen 1/10 second apart. This is more difficult to discern for people with schizophrenia, suggesting they might live in an expanded moment and have trouble dividing that moment below a certain level of scale. This might explain a variety of symptoms. A number of recent studies have shown that altering ones' experience of time can induce schizophrenia-like symptoms.

One example used healthy volunteers to play a video game steering a plane around obstacles. When the experimenters introduced a .2-second delay between the players' action and the games response, the players learned to correct for this, eventually perceiving their action and the response as simultaneous. Intriguingly, when the experimenters removed the .2-second delay, the players felt as if the game anticipated their actions. If one were to experience other parts of their life in this manner, they might feel externally controlled, as if by another person, or perhaps that the T.V. was broadcasting their thoughts, or there was an organized conspiracy against them, much like the experiences of people with schizophrenia.[1]

It may seem that in this situation time has not sped up or slowed down, but that it is out of phase (the frequencies do not line up). I want to draw attention to the fact that time dilation can appear as a phase shift. Consider the twin paradox. One twin goes on a journey at speeds close to the speed of light and returns younger than the twin who stayed behind on the stationary planet. If one did not consider the underlying time dilation, it might appear that their times were simply misaligned.

It seems to me that the video game players compensated for the imposed delay by shifting their internal time ahead to align with the game, but that when the delay vanished it seemed to the players that the game had

---

[1] (Fox 2009)

shifted back in time creating the full effect of time dilation rather than just a phase shift.

The fact that lack of dopamine is associated with general errors in time estimation rather than only with mental acceleration or slowing of external time, lends credence to the hypothesis of Donald Woodward that dopamine is not necessarily the clock itself, but perhaps just the clock-watcher. He proposes that the basal ganglia provide focus, and sometimes focuses on time. Meck's studies involved brain scans of people tracking time, so it seems possible that the scans were picking up the neural signature of "tracking" rather than the neural signature "time."[1] Meck has since incorporated this perspective into his theory, recognizing that basal ganglia, with the help of dopamine, seem to track activities of the frontal cortex, possibly deriving its sense of time from the rate of frontal cortex activity.[2]

I suggest that dopamine does play a role in facilitating our ability to focus and that it is precisely our focus, or attention, that affects our perception of time. Dopamine, the mental accelerator, is produced when we are frightened, alarmed, stressed, or deeply concentrating. All of these situations are synonymous with mental acceleration and the need to pay attention.

In fact, it seems that surprise activates the dopamine system. When we receive expected rewards, the dopamine system, and thus, the perceived rate of time, is unaffected. Unexpected rewards, on the other hand, generate the release of dopamine, and the lack of an expected award causes a decline in dopamine release. Both changes in dopamine levels signal that something has deviated from expectation and that we should pay attention.

Thus, when teaching a new skill, maximum learning occurs through rewards distributed sporadically, rather than every time the new skill is demonstrated. Maintaining the element of surprise keeps the dopamine system engaged, and thus keeps the student engaged. The entrainment generated by random rewards is part of the reason that gambling and checking your phone, is so addictive. If the rewards are too regularly distributed, they fail to generate the level of desire that unexpected rewards generate.[3]

This makes sense because desire is based on not having. If you are guaranteed the reward, you basically already have it, so there is no motivation

---

[1] (McCrone 1997)
[2] (Williams 2006)
[3] (Blakeslee 2002)

to put forth any extra effort to achieve it. It is, in effect, the motivation of risk and challenge that we crave, not necessarily the prize, because it is through that motivation that we feel that we are operating at our peak capacity.

When we are paying attention to something, we are not paying attention to time and it seems to pass quickly. When nothing holds our attention, we become bored. Then, we pay attention to time and it seems unbearably slow. The dopamine system not only perks us up to get us to pay attention and learn something, it also conditions us to seek out the pleasurable engagement of challenges, risk taking, and the element of surprise. If attention is linked to the synchronization of time scales, as I suggest, then too little or too much dopamine prevents that synchronization, making it difficult to relate individual desires to the requirements of the external world.

4.4 Intention

I have explored the role of repetition in creating time, and of attention in shifting our perception of time based on synchronization of time scales and desire. Now I take that one step further to consider the role of intention. The repetition of wave function collapse builds temporal depth and progression, both in the external world and as mirrored in consciousness. Attention synchronizes the frequencies of the brain with external frequencies in order to absorb information and relate to external reality. Intention then drives brain frequency to entrain external reality to its frequencies. Anytime a person's activities deviate from the surrounding frequencies, like changing the drumbeat in a drumming circle, his or her intention manifests, seeking followers--no longer content to follow. Attention, aligning one's frequencies with external frequencies, can also be intentional–thus purposefully aligning one's brain waves with external frequencies.

I propose that consciousness functions optimally when intention and attention balance one another in an ever refining back and forth dance, creating a continually deepening boundary between receptivity and contribution from which complexity unfurls. Intention relies both on our ability to remember, to predict the future, and to exert control over our actions. Repetition and attention are, in a sense, value free. While selective attention does filter reality, it is intention, whether conscious or unconscious, that directs the attention, imposes its agenda, and wields the powerful influence of desire on our perception of time.

According to Vrobel a richer experience, containing more levels of description, will pass more quickly than one containing fewer levels of description. It is logical then that time seems to pass more quickly the older we get since we have a larger knowledge, memory, and association store to enrich every experience. This issue undergirds the modern person's struggle with the paradox of time. People tend to feel they do not have enough time, as if it is passing them by too quickly. If this perception of accelerated time has to do with the richness of the experience and you want to slow your experience of time down, simply stop doing things and clear your mind. Do not divide your day into as many little parts. Then the problem becomes potential boredom or lack of stimulation. This is where the perspective of fractal optimization might be helpful in gaining a deeper understanding of the dynamic balance between scheduled and free time, receptivity and contribution, desire and acceptance.

4.4.1 Desired Time

I characterize the influence of intention on subjective time by the general rule that desired time is inversely proportional to experienced time. The more time you want, the faster it seems to evaporate. The faster you want something to pass, the longer it takes, the more you want it. Desire more often leads to its own re-entrenchment rather than to its satisfaction. Our very attempt to control time, to conform its behavior to our desires, seems to backfire.

I remember a specific childhood experience, watching the clock and waiting for my grandparents to arrive. The more I wanted them to get there, the more often I checked the clock, the more infuriating was the slowness of the clock, and the more I wanted them to arrive! I was stuck in an emotional spiral of desire that bred frustration. Of course, I quickly realized I would have to focus on something else if I wanted the time to go faster. Often, we forget these simple lessons, as we grow older and think ourselves in circles of complexity justifying our frustration and desire with the help of a society who encourages our self-indulgence.

If the more time we want, the less time we have, how can we possibly work our experience of time to our advantage? This is a point of reversal similar to the uncertainty principle, revealing the limits of what we can control. We must learn to work within its bounds.

It is appropriate that time is said to flow like water because, like water, one must work with it rather than try to conquer it. Like water, time reflects back to you what you present to it. If you have ever been under water, you know its power to silence, support, and balance. It lets you feel like you are flying; but if you fight against it, it will work against you. Water reflects and magnifies not only physical objects, but also reflects and magnifies your fears and your peace, respectively. What you put in you get out. Pollution will come back to haunt your drinking water. Fear will drown you. Water can show you most immediately how becoming one with something outside of your body will result in an expansion of the powers of self through the expansion of self.

In the same way, we can see time as an oppressive or a supporting force, as a teacher or as an enemy. Intentions make all the difference in whether you sink or swim, the difference between chaos and sanity.

4.4.2   Dopamine and Motivation

Fascinatingly, just as attention and intention complexly intertwine in theory, so they intertwine biochemically in the brain, both in intimate relationship with dopamine. As discussed previously, dopamine plays an important role in our attention and sense of time. Alteration in the production or utilization of dopamine prevents our ability to synchronize our inner clocks with external frequencies. This asynchrony leads to difficulties in maintaining attentiveness and/or relating to others whose timescales dramatically differ from our own.

In this section, I explore another side of dopamine, specifically the role it plays in desire and motivation and how this affects our perception of time. It seems that dopamine may more closely connect with anticipatory desire than with the satisfaction of that desire.

Dopamine, which is associated with sexual arousal/motivation, inhibits another neurotransmitter, prolactin, which is associated with sexual gratification.[1] Prolactin, in turn, inhibits dopamine–hence, sexual arousal is more difficult after sexual gratification.[2] These neurotransmitter correlates for desire and satisfaction play a role to our temporal experience. If

---

[1] (Ben-Jonathan and Hnasko 2001)
[2] (Brody and Krüger 2006)

dopamine is the neurotransmitter associated with our perception of time, then prolactin is the neurotransmitter associated with our experience of timelessness, and the two alternate through mutual inhibition in the same way we alternate between desire and satisfaction, time and timelessness.

During satisfaction, people experience timelessness. Timelessness is characterized by lack of dopamine as seen in Sacks' patients, and as experienced naturally during the prolactin release of breastfeeding and orgasm. Notably, prolactin also inhibits menstruation in women. Menstruation has been postulated as one of the first reasons to mark time cycles longer than the day/night cycle in the evolution of consciousness.

It is also interesting that prolactin stimulates the proliferation of cells that eventually become the axon's lipid coating, speeding signal transfers between neurons.[1] These signals travel through changing protein conformation within the neuron. These are the same proteins that Penrose and Hameroff postulate participate in quantum superposition, and that I suggest touch timelessness while superposed. Recall that hydrophobicity plays a key role in the preservation of this superposition. Thus, the lipid layer that prolactin helps create may also contribute to the experience of timelessness in consciousness.

The timelessness of satisfaction, however, inevitably passes due to some change that leads to desire and temporality once again. Desire can make time seem faster or slower depending on whether you thrive on the state of arousal to action or wallow in the frustration of not having that which you desire. If you have ever experienced an intense desire for something, whether it be a thick, moist slab of chocolate cake or your latest crush, you know how fast your brain can spin trying to come up with a way to get (or to rationalize getting) what you want. As I illustrated in a previous section, dopamine's association with desire links this brain spinning to increased frequency of neural cycling.[2] The more our brains spin trying to find a way to bring the object of our desire closer, the more out of sync we become with external time.

If our love is requited, or the chocolate cake yields itself to our desires, or we feel confident in our possession of it, we settle into the bliss of satisfaction and seeming timelessness that emerges during contentment. All scheming can go out the window, our brains can relax and enjoy. However, the pleasure is remarkably fleeting. Time spent in contentment

---

[1] (Gregg et al. 2007)
[2] (McCrone 1997)

quickly vanishes because of its lack of division. When the brain slows down, external time seems to pick up the pace, racing by before we get a chance to enjoy it.

Before you know it, you are at the mercy of desire again. The cake is gone, the thrill of unknowing over, sets the stage for the next desire to take hold. A spinning brain is like a spinning tire, digging deeper into time instead of sand. A spinning brain creates more content, more divisions of time, more levels of description, and increases the density of time as you deepen into it. Desire divides the timelessness of the moment, creating content, momentum, and time from the longing of separation.

If the object of our desire is out of reach, the increased density of thought, can inadvertently slow our perception of external time, thus seemingly prolonging the suffering of our separation from that object of desire. However, the timelessness of satisfaction can also co-exist with temporal desire. If you can stay in the moment, you can stay in the timelessness of satisfaction. Only when you check the clock does time seems to move slowly. Here attention and intention play a pivotal role. If you allow yourself to enjoy the process as much as you would the prize, maintaining focus on the task at hand and not continually seeking for that which is not present, then you can maintain a sense of cooperation with time rather than antagonism.

If you happen to love the infatuation and the chase of love, then even when time marches painfully slow in the uncertainty of your love's return, you might relish that painful slowness of separation. You might feel enlivened by the challenge of winning an uncertain prize, thus spinning in a flow state of creative timelessness devising romantic seductions. Thus, the temporal effects of desire can be paradoxical, slowing or speeding one's perception of time based on where one places their attention.

If your brain spins on the subject of separation, you will dig yourself deeper into separation, feeling more separate. If your brain spins on bringing the object of your desire closer, you will feel closer to that object, closer to satisfaction and closer to timelessness. However, deeper into yourself is not closer to an external object of desire, so this can generate a painful paradox when confronted with external reality.

We love the forgetting of time that comes with the flow of an experience that challenges us just the right amount,[1] but have a hard time with the fact that all our scheming does not always bring us to our goal. In

---

[1] (Csikszentmihalyi 1990)

fact, this potential for failure provides great motivation. Without the potential for failure, we would not try nearly as hard or even at all. The timeless absorption in a task breaks upon comparison with external time. This is similar to the timelessness of superposition that then collapses into temporality upon measurement. In this case, the experience of timelessness collapses with the measurement of checking the clock, or the location of one's desire, and comparing expectation with reality. When a speedy brain looks at a clock, it might realize that barely any time has passed compared to how many circles, layers, and thoughts it has covered.

The fact that dopamine links our attentiveness to novelty and creates desire, can go two different ways depending on whether we focus on the journey or on the separation. It can drive us to seek out challenges and creatively tackle them, or we can become trapped in a tightening spiral of addiction. Earlier I used the example of the addiction of phone checking. We get so easily stuck in this addiction because it serves the dopamine system, we are sporadically rewarded with social contact and thus learn the behavior and seek it out repeatedly hoping for the reward of more social contact.

Unfortunately, the good old advice, "fake it till you make it" works equally well positively or negatively. The more you check your phone looking for social contact the less contacts you will seem to receive, since the frequency of reward per measurement declines as the frequency of seeking it increases. Thus, by seeking social contact, you reaffirm to yourself that it is something you are lacking. The more diligently you seek this, the more absent it feels, even if your social calendar is in fact booked. On the other hand, if you focus on the abundance of contacts that you receive, then you might reason that you really cannot possibly check your phone because of the vast amounts of time it would take up. Then you will put off the checking and once again you have created a self-fulfilling prophecy of massive amounts of contacts, without changing the facts of the situation. Obviously, if you check your phone only once a day then you will receive more contacts per check than if you check fifty times a day.

As Alan Watts puts it, "The moment we look for union with God we imply that we do not already have it."[1] Seeking, measuring, checking all focus on the lack and become self-fulfilling. Checking the clock focuses on when something will be over or when something else will come; it takes you out of the present and becomes a self-fulfilling prophecy of lacking time.

---

[1] (Watts 1947, 94)

Taking a quantum measurement also focuses on separation and becomes a self-fulfilling prophecy of particulate isolation. Seeking continual reassurance in a relationship focuses on the separation and can drive the other person away. By questioning your happiness, you introduce doubt, often thereby spoiling the subtle contentment of happiness. Sometimes it is preferable not to collapse the wave function or time, love, or happiness, but to allow it to live and work its magic without our interference.

4.4.3 Grasping/Uncertainty

We have managed to domesticate and control space to serve our purposes, at least on a small level. In the case of time, however, we focus on its measurability precisely because our experience of time is so slippery and uncontrollable. Time continually evades our grasp and our desires for its behavior. It remains wild. As Brian Swimme and Thomas Berry tell us, "A *wild* is a great beauty that seethes with intelligence that is ever surprising and refreshing for the human mind to behold."[1] Time laughs at our futile attempts to stop it and dawdles when we would have it fly.

Our discontent with time manifests in two complimentary ways. When life gives us repetition, we want novelty. When life gives us change, we want stability. A materialistic lifestyle is a manifestation of both of these discontents, offering the endless novelty of a throwaway society and the solid stability of material form. However, as with most of our efforts to control time, materialism has unintended consequences.

The need to counterbalance flux with a certain amount of stability can sometimes lead to dogmatism, fear of otherness, closed mindedness, boredom, and materialism. In our resentment of time's freedom, we cling to stability wherever we can find it. Perhaps we feel that the more things we own, the more time we have managed to preserve in nice little packages, or the more things we own the more time they will save us.

The more time we "save," the less time we have, because saving time means increasing the temporal density by fitting more processes into a smaller amount of time, which generates new LODs (levels of description), increasing the temporal depth and contracting the moments of time generating an appearance of accelerated time.

As the pace of life quickens everything–from advertising to children–must compete for attention. In the dopamine system, the element

---

[1] (Swimme & Berry 1994, 127)

of surprise is especially effective at garnering not only attention but establishing intention and repetition. Because of this, our culture breeds novelty and stimulation through channel surfing, consumerism, one-night stands, overly dramatized media, and travel.

We tend to seek stability in materiality without cultivating the stability of the self, and flux in our material surroundings without cultivating our inner creativity. An increased awareness of how stability and flux evoke one another in an ever-tightening spiral allows us to take some agency in how this dynamic plays out in our lives, rather than just perpetuating our societal entrenchment in unsustainable consumption. We can choose to step off the ever-accelerating trajectory of skimming across the surface of modern life and sink deeply into the timeless depths of the moment, cultivating the sustainability of our internal creativity and integrity as our personal balance between flux and stasis.

If consciousness and science seek to define time as an object, they miss the essential nature of the flux of time. If the mode of observation is the problem, if we cannot come to know time's flowing nature by imposing stasis on it, what method of observation might we employ to gain an accurate description of and a more satisfying relationship with time?

One place internal stability and creativity can be found is in contemplation, reflection, and meditation—stilling one's mind, body, or seeking quiet surroundings. The modern secular world has largely failed to encourage people to carve out niches for the practice of contemplation and reflection, with the exception of pockets where spiritual traditions maintain this practice.

The practice of contemplation, reflection, or meditation not only provides undivided time to contrast with busy-ness, but also allows the space for the self to have a relationship with itself, deepening its internal resources for dealing with the flux of the external world. The boundary between the self and the external world is a continually deepening fractal boundary, which maintains its momentum and definition through the full and equal participation of each member. If the self is not deepening its own duality through balancing activity with stillness and observation with creative contribution, then not only does the individual experience imbalance internally, but the relationship with the external world also becomes unbalanced. Spacetime holds apart the localized self and its larger environment in dynamic tension and provides the channels for the flow of their mutual exploration.

The use of meditation to come to know a flowing mind and reality offers an alternative way of knowing. This way of knowing complements dominant ways of knowing, such as logic, science, and definition by objectification. A fractal model of time exemplifies this principle of balancing both stasis and flux through a model that holds the paradox of infinity and finiteness. As a fractal illustrates infinite surface area within a finite space, so does a finite moment hold within it the infinity of eternity, and likewise the stillness of eternity contains within it the infinite flux of the multitude of moments. We can experience this paradox through the kōan of a fractal or through our own experience in meditation.

To observe the mind, we observe thought using thought itself. This only serves to complexify that which we hoped to simplify, by adding an element of nonlinear feedback. Instead of feeling thwarted, perhaps we can appreciate this as a system of reiterative feedback that keeps us poised on the growing edge of life's complex unfolding. The reiterations, putting the output back in as the input, also builds fractals. The more closely you try to define the boundary of a duality, the more complexity unfolds.

Meditation, on the other hand, takes a slightly different approach. Instead of chasing thought in an attempt to drive it to some sort of definition, the idea, paradoxically, is to quit chasing the thoughts. Let thoughts arise and slide away, identifying with the still point of observation, rather than the endless action of the monkey mind. Thus, instead of trying to still the flow of thoughts as an object for observation and definition, the subject itself is stilled, achieving clarity by identifying with stillness, rather than by stilling motion.

In order to see, define, or know something, it is often necessary to separate one's self from it. When one identifies with stillness, one can see motion. When one identifies with motion, to define something else that is also in motion, one must impose stillness on it. The kind of understanding the mind achieves in the cessation of thought through meditation is not causal; it is more akin to Penrose's *insight,* suspending the collapse of the wave function that occurs with every thought, in order to contact timelessness. It is eternally present, but only attainable through seemingly paradoxical means. One cannot achieve it through the desire to achieve it. One must first unconditionally accept his or her present state and simultaneously accept any change that might occur.

We try to grasp time with our minds, like we try to grasp thoughts, like we try to grasp life, only to find that grasping reveals our false belief that we are not already complete and capture denies the essential freedom of both ourselves and of the object of our desire.

Is the answer to peace with time a matter of controlling, not time, but our desires, as the Buddha would prescribe? I do not intend to preach an ascetic denial of desire, because there is fun in the game of grasping in and of itself, which animates life. Through the recognition of the limits of our grasp, and of self-denial, we may actually free ourselves from capture of the inadequacy of our own concepts. The goal is not the goal; it is a means to the process. Once again, we find ourselves face to face with cyclicity and paradox; the goal is not the goal; the goal becomes the process and the process becomes the goal. We do not speak and write in order to define everything explicitly in word and writing. We speak and write to facilitate the process of communication. It is this point of reversal that I want to emphasize in our relationship with time.

4.4.4 Co-Creating Time

Today, when one thinks of time, one rarely considers their subjective experience. More often one thinks of the clock and how there is never enough time. Modern humans allow themselves to become slaves to the clock and to the demands of outer experience and interactions. This creates the texture of skimming across the surface of life driven from one event to the next by our master the clock, never allowed to sink into the depths of a moment. When scheduled to the brim, we hand authority over to the clock, often forgetting how to maintain our own centered-ness and self-authority within that schedule.

This dynamic of clock domination is characteristic of the imbalance that can come with either/or thinking. The demand for an abstract "either/or" ordering of priorities that extends across all areas of our life often prevents the organic unfolding of complex "both/and" specificity within each unique context as it arises. Sometimes it is beneficial to abide by the structure provided by the clock, but we also need to balance that movement through time with stillness within it, to be present to the present in order to empower ourselves. One may forget their power to co-create time.

The dominion of quantization, mathematics, physics, and patriarchy colonized time. Nature's clock used to rule, offering much greater leeway. Temporal demands were seasonal (when to plant, when to harvest) and daily (when to sleep, when to wake). These external demands have specific ranges in which they preside, and the rest of time was left largely to divisions imposed by society or by individual choice. One can recapture this feeling of living on nature's clock by immersing oneself in wilderness. As John Muir put it, "Life seems neither long nor short, and we take no more heed to save time or make haste than do the trees and stars. This is true freedom, a good practical sort of immortality."[1]

In modern life, however, "freed" from the demands of natural temporal rhythms, the unyielding rule of the clock takes the throne. Time was quantized long before the clock, but it was a large-scale quantization that left lots of room for internal creative contribution rather than mere reaction. One could divide the "in between" time as one liked, thus enriching the quality of time within the quantified chunks.

The small-scale quantization that came with the clock leaves much less room for personal agency. One great irony of our culture of so-called "independence" is that it demands such conformity. The subjective experience or "opinion" holds no weight against objective fact. Facts, however, are abstractions and thus can never capture the complexity and specificity of truth in the way available to a single perceiving subject.

We found power in the ability to orchestrate our movements with the help of the clock. Physics predicts movement in time. It reveals the future to a precision otherwise unavailable. Objective time is regarded as real, true, and consequential. Our subjective experiences of time have been relegated to the realm of illusion, triviality, and inconsequence, along with religion and arts, which facilitate a deeper experience of time.

Consider time as omnigenetic—always beginning—with every beginning. Each lifetime, blinking in and out of existence, from a particle's blip to the eons of star's, contain within them the beginning, the end, and the eternity of time, just as much as the entire lifetime of the universe does. These time scales nest within one another.

Recognizing time's omnigenetic nature, one might more fully embrace the role of co-creators of temporal reality. Each moment begins anew. Intentions, however, do not always yield predictable results. When exercising intention, desire and grasping can actually thwart our true aims. Often a desire for material things hides a desire for appreciation and connection, the very things greed might drive away.

---

[1] (Muir 1911, 52)

Yet wielding intention in a non-grasping spirit—by allowing, appreciating, and inviting--we gain participation with the mystery–with time, nature, loved ones, and ourselves. As co-creators of time, may we bestow upon our creations that which we hope to receive from our creators: love.

As for me, I try to increase my "time savoring" over my "time saving." We have technologies to assist us: wilderness, meditation, contemplation, bodily awareness, story, prayer, and reflection. Whatever adds layers of meaning to our experiences, increases each moment's depth, and thus the gravitational pull to keep us firmly rooted in the unfurling present. I suspect that our next era might rely more heavily on recalling the spiritual technologies of the past, and creating new technologies of spirit, aesthetics, and symbolism to increase the depth and richness of our experiences, deepening the quality of time rather than worrying about its quantity.

The theory of fractal time presented in this book seeks its integral nature by preserving essential paradoxical mystery at many turns, an in particular by holding a place for both physical and subjective theories of time.

The Mandelbrot set's vivid contrasts and textures unfurl from a finite line, from an austere mathematical equation, inspiring awe. The infinite within the finite–this is the paradox that animates the world–eternity within a moment, the moment within eternity, and the whole body of the universe in between, chasing its tail. We approach and retreat, buffeted by the unfurling waves, flirting with the dissolution of everything into unity, yet decide to hold the universe open, for just one moment longer.

Figure 4.3 Ouroborus[1]

---

[1] Pelecanos, Theodorus. 1478. Fol. 279 of Codex Parisinus graecus 2327, a copy an early medieval tract attributed to Synosius of Cyrene (d. 412). Text attributed to Stephanus of Alexandria (7th century).

References

Abbot, Edwin Abbott. 1884. *Flatland*. London: Seeley. http://www.archive.org/stream/flatlandromanceo00abbouoft#page/n5/mode/2up (accessed February 27, 2010).

Aharonov, Y., and L. Vaidman. 1990. Properties of a quantum system during the time interval between measurements. *Physical Review*. A41:11.

Aristotle. 350 BCE. *Physics*. Trans. R. P. Hardie and R. K. Gaye. Internet Classics Archive http://classics.mit.edu/Aristotle/physics.2.ii.html (accessed Sept 8, 2008).

Ben-Jonathan N, Hnasko R. 2001. Dopamine as a Prolactin (PRL) Inhibitor. *Endocrine Reviews* 22 (6): 724–763. http://edrv.endojournals.org/cgi/reprint/22/6/724.pdf (accessed March 31, 2010)

Bell, J.S 1964. On the Einstein Podolsky Rosen Paradox, *Physics 1*, 195-200. http://www.drchinese.com/David/Bell_Compact.pdf (accessed March 5, 2010)

Bergson, Henri, 1946. Time is the flux of duration. *The creative mind: An introduction to metaphysics*. Trans. Mabelle L. Andison. New York: Philosophical Library.

Bierman, Dick J. 2002. An fMRI study of anomalous anticipation of emotional stimuli. Paper presented at the Towards a Science of Consciousness conference, April 10, Tucson, AZ. http://www.quantumconsciousness.org/ppts/tucson-2002.Ppt#257,2 (accessed February 27, 2010).

Beirman, Dick, and Steven Scholte. 2002. Anomalous anticipatory brain activation preceding exposure of emotional and neutral pictures. http://www.quantumconsciousness.org/pdfs/presentiment.pdf (accessed February 27, 2010).

Beyer, Wolfgang. 2019. Mandelbrot set, *Wikipedia, the free encyclopedia*, https://commons.wikimedia.org/wiki/User:Wolfgangbeyer (accessed June 4, 2019).

Blakeslee, Sandra. 2002. Hijacking the brain circuits with a nickel slot machine. *New York Times*, February 19. http://www.nytimes.com/2002/02/19/science/hijacking-the-brain-circuits-with-a-nickel-slot-machine.html (accessed September 22, 2009).

Bohm, David. 1951. *Quantum theory*. New York: Prentice Hall.

Bohm, D. 1986. Time, the implicate order, and pre-space. In *Physics and the ultimate significance of time*, ed. D. R. Griffin, 177-208. Albany: State University of New York Press.

Bohm, David. 1990. A new theory of the relationship of mind and matter. *Philosophical Psychology*, Vol. 3, No. 2, 271-286.

Bohr, N. 1934. *Atomic theory and the description of nature*. Cambridge UK: Cambridge University Press.
Bothun, Greg, ed. 2000. Astronomy 123: Cosmology and the origins of life. http://zebu.uoregon.edu/hb/2.10.jpg (accessed February 8, 2010).
Brody, S., and T.H.C. Krüger. 2006. The post-orgasmic prolactin increase following intercourse is greater than following masturbation and suggests greater satiety. Biological Psychology, 71, 312-315.
Brookhaven National Laboratory. 2001. New Brookhaven lab study shows how Ritalin works. Jan. 16, 1991. http://www.bnl.gov/bnlweb/pubaf/pr/2001/bnlpr011501a.html (accessed August 31, 2009).
Buccheri, R., M. Saniga, and W. M. Stuckey, eds. 2003. *The nature of time: Geometry, physics and perception*. Dordrecht: Kluwer Academic.
Buchanan, Mark. 2009. Can fractals make sense of the quantum world? *New Scientist* 2701 (March 30), http://www.newscientist.com/article/mg20127011.600-can-fractals-make-sense -of-the-quantum-world.html?full=true (accessed February 27, 2010).
Burnham, Douglas. 2006. s.v. Gottfried Wilhelm Leibniz. Internet Encyclopedia of Philosophy. http://www.iep.utm.edu/l/leib-met.htm (accessed November 16, 2006).
Coghlan, Andy. 2009. Time moves too slowly for hyperactive boys. *New Scientist* 2711 (June 10), http://www.newscientist.com/article/mg20227115.100-time-moves -too-slowly-for-hyperactive-boys.html?DCMP=OTC-rss&nsref=online-news (accessed August 3, 2009).
Costa de Beauregard, O. 1989. *Bell's theorem, quantum theory, and conceptions of the universe*. Ed. M. Kafatos. Dordrecht: Kluwer.
Csikszentmihalyi, Mihaly. 1990. *Flow: The psychology of optimal experience*. New York. HarperCollins.
Davies, Paul. 1995. *About time: Einstein's unfinished revolution*. New York: Touchstone.
Edinger, Edward F. 1984. *The creation of consciousness: Jung's myth for the modern man*. Toronto, Canada: Inner City Books.
Elitzur, A. C., and L. Vaidman. 1993. Quantum-mechanical interaction-free measurements. *Foundations of physics*. 23:987-97.
Einstein, A., B. Poldosky, and N. Rosen. 1935. Can quantum –mechanical description of physical reality be considered complete? *Physical review*. Vol 47. 777-780.
Fox, Douglas. 2009. Timewarp: How your brain creates the fourth dimension. *New Scientist* (October 21) 2731, 32-37.
Gebser, Jean. 1986. *The ever-present origin*. Trans. Noel Barstad and Algis Mickunas Athens: Ohio University Press.
Gleick, J. 1987. *Chaos: Making a new science*. New York: Viking.
Gödel, Kurt. 1949. An example of a new type of cosmological solution of Einstein's field equations of gravitation. *Reviews of Modern Physics*. 21:447.
Greene, Brian. 2000. *The elegant universe*. New York: W. W. Norton.
Gregg C, Shikar V, Larsen P, et al. 2007. White matter plasticity and enhanced remyelination in the maternal CNS. *J. Neuroscience*. **27** (8): 1812–23.
Grush, Rick, and Patricia Chuchland.1995. Gaps in Penrose's tilings. *Journal of Consciousness Studies* 2 (1). http://mind.ucsd.edu/papers/penrose/penrosehtml/penrose-

text.html (accessed January 29, 2010).
Haramein, Nassim. 2004. *Crossing the event horizon*. VHS. Workshop proceedings October, Ashland, Oregon.
Hameroff, Stuart. 1999. "Quantum mechanics in the brain?" Quantum approaches to understanding the mind. University of Arizona webcourse. Nov. 2, 1999.
Hameroff, Stuart. 2003. Time, consciousness and quantum events in fundamental spacetime geometry. *The nature of time: Geometry, physics and perception*. Eds. Buccheri, R., M. Saniga, and W. M. Stuckey. Dordrecht: Kluwer Academic.
Hameroff, Stuart. 2006. The entwined mysteries of anesthesia and consciousness. *Anesthesiology* 105:400-412.
http://www.quantumconsciousness.org/documents/twined_000.pdf (accessed January 29, 2010).
Hameroff, Stuart. 2009a. What is consciousness? www.quantumconsciousness.org/presentations/whatisconsciousness.html (accessed February 3, 2009).
Hameroff, Stuart. 2009b. Breakthrough study on EEG of meditation. http://www.quantumconsciousness.org/EEGmeditation.htm (accessed Feb. 11, 2009).
Hameroff, Stuart, and Roger Penrose. 2009. Conscious events as orchestrated spacetime selections. http://www.quantumconsciousness.org/penrose-hameroff/consciousevents.html (accessed March 31, 2009).
Hawking, Stephen, and George Ellis. 1973. *The large scale structure of space-time*. Cambridge, UK: Cambridge University Press.
Hebert, Nick. 1985. *Quantum reality*. New York, New York: Doubleday.
Hoagland, Hudson. 1981. Some biological considerations of time. *The voices of time: A cooperative survey of man's views of time as expressed by the science and humanities*. ed. J.T. Fraser. Amherst: University of Massachusetts Press.
Hosinski, Thomas. 1993. *Stubborn fact and creative advance: An introduction to the metaphysics of Alfred North Whitehead*. Lanham, MD: Rowman & Littlefield.
Irigaray, Luce. 1985. When our lips speak together. *This sex which is not one*. Trans. Catherine Porter. Ithaca, NY: Cornell University Press.
Irigaray, Luce. 1993. Sexual difference. *An ethics of sexual difference*. Trans. Carolyn Burke and Gillian C. Gill. Ithaca, NY: Cornell University Press.
Jaroszkiewicz G. 2003. Analysis of the Relationship Between Real and Imaginary time in Physics. In: Buccheri R., Saniga M., Stuckey W.M. (eds) *The Nature of Time: Geometry, Physics and Perception*. NATO Science Series (Series II: Mathematics, Physics and Chemistry), vol 95. Springer, Dordrecht
Jaroszkiewicz G. 2016. *Images of Time: Mind Science and Reality*. Oxford University Press. Oxford, UK.
Keim, Brandon. 2008. Brain scanners can see your decisions before you make them. *Wired* (April 13)
http://www.wired.com/science/discoveries/news/2008/04/mind_decision (accessed 2/28/10).
Kelly, Sean. 2008. Integral time and the varieties of post-mortem survival. *Integral Review* (June) 4 (1). http://integral-review.org/current_issue/index.asp (accessed 2/28/10).

LeFevre, Eric. 2009. Scale relativity. http://luth2.obspm.fr/~luthier/nottale/ukrechel.htm (accessed 2/28/10).
Marek-Crnjac, L. 2009. A short history of fractal-Cantorian space-time. *Chaos, Solitons and Fractals 41 2697–2705*
McCrone, John. 1997. When a second lasts forever. *New Scientist* (November 1). http://www.newscientist.com/article/mg15621065.300-when-a-second-lasts-forever--its-not-just-in-the-movies-that-moments-of-crisis-seem-to-pass-in-slow-motion-john-mccrone-investigates-the-tricks-our-minds-play-with-time.html (accessed 2/28/10).
McTaggart, J. E. 1908. The unreality of time. *Mind. A Quarterly Review of Psychology and Philosophy*. Ed. G. F. Stout.
McTaggart, John M. E. 1921. Time is not real. *The nature of existence*. Ed. C. D. Broad. Cambridge, UK: Cambrige University Press.
Morin, Edgar. 1977. *La Méthode I: la Nature de la Nature*. Paris: Seuil.
Morin, E. 1980. *La Méthode II: la Vie de la Vie*. Paris: Seuil.
Morin, Edgar. 1981. *Pour Sortir du XXe Siècle*. Paris: Fernand Nathan.
Muir, John. 1911. *My first summer in the sierra*. New York: Houghton Mifflin. http://nature.gardenweb.com/muir/firstsummer (accessed February 5, 2010).
Nahin, Paul J. n.d. *An Imaginary Tale: The Story of the Square Root of -1*. Princeton, NJ: Princeton University Press.
Penrose, R. 1994. *Shadows of the Mind: A Search for the Missing Science of Consciousness*. New York: Oxford University Press.
Sacks, Oliver. 1999. *Awakenings*. New York: Vintage Books.
Newton-Smith, W. H. 1980. *The structure of time*. London: Routledge & Kegan Paul.
Nottale, Laurent. 1993. *Fractal spacetime and microphysics: Towards a theory of scale relativity*. River Edge, New Jersey: World Scientific.
Penrose, R. 1989. *Emperor's new mind: Concerning computers, minds, and the laws of physics*. New York: Oxford University Press.
Penrose, R. 1994. *Shadows of the mind: A search for the missing science of consciousness*. New York: Oxford University Press.
Penrose, R. 2005. *The road to reality: A complete guide to the laws of the universe*. New York: Alfred A. Knopf.
Plato. 5th century BCE. Trans. B. Jowett. Time. *Timaeus*. Greek Word Library http://www.ellopos.net/elpenor/physis/plato-timaeus/time.asp (accessed 2/28/10).
Pribram, Karl, and Deborah Rozman. 1997. *Early childhood development and learning: What new research on the heart and brain tell us about our youngest children*. Paper presented at White House Conference on Early Childhood Development and Learning, April 17, San Francisco, CA. http://www.thesecularspirit.com/text/heartbrain.htm (accessed 9/10/09).
Prigogne, I., and Stengers, I. 1984. *Order out of chaos: Man's new dialogue with nature*. New York: Bantam Books.
Sacks, Oliver. 2004. Speed. *New Yorker*, August 23.
Sacks, Oliver. 1973. *Awakenings*. New York: Vintage Books.
Saniga, Metod. 2003. Geometry of time and dimensionality of space. *The nature of time: Geometry, physics and perception*. Eds. Buccheri, R., M. Saniga, and W. M. Stuckey.

Dordrecht: Kluwer Academic.
Scott, Alex. 2002. Henri Bergson's *The Creative Mind*.
http://www.angelfire.com/md2/timewarp/bergson.html (accessed 2/28/10).
Sherover, Charles M., ed. 2001. *The human experience of time: The development of its philosophic meaning*. Evanston, IL: Northwestern University Press.
Swimme, Brian, and Thomas Berry. 1994. *The universe story*. New York: Harper Collins.
Tarnas, Richard. 1991. *The passion of the Western mind*. New York: Ballantine Books.
Vrobel, Susie. 1998. *Fractal time*. Houston, TX: Institute for Advanced Interdisciplinary Research.
Watts, Alan. 1947. *Behold the spirit: A study in the necessity of a mystical religion*. New York: Random House.
Whitehead, Alfred North. 1929. *Process and reality: An essay in cosmology*. New York: Harper and Brothers.
Whitehead, Alfred North. 1933. *Adventure of ideas*. New York: Free Press.
Whitrow, G. J. 1980. *The natural philosophy of time*. 2nd ed. Oxford, UK: Clarendon.
Werbos, P. 1989. Bell's theorem: The forgotten loophole and how to exploit it. In *Bell's theorem, quantum theory, and conceptions of the universe*. Ed. M. Kafatos. Dordrecht, Netherlands: Kluwer
Williams, Caroline. 2006. Teach your brain to stretch time. *New Scientist* (February 4). http://www.newscientist.com/article/mg18925371.700-teach-your-brain-to-stretch-time.html?full=true (accessed 2/28/10).

Index

40 Hz, 19, 116
A series, 87
abstraction, 104–6, 28, 85
acausal, 47, 105, 106
actual occasion, 73, 104, 106
Adderall, 120
ADHD, 120
Alpha waves, 116
amphetamine, 121
amplitude, 14, 71, 77, 80, 113
anesthesia, 7, 8, 9, 15, 19, 21, 137
antiparticle, 33, 72, 74, 103
Archimedes, 9
arrow of time, 25, 31, 80
asynchrony, 125
attention, 129–30, 116–24, 112, 24, 30, 53, 73, 90
awareness, 109–11, 21, 43, 51, 115, 120, 130, 134
B Series, 87
background radiation, 103
backwards time, 31, 32, 34, 50, 74
basal ganglia, 90, 119, 120, 122
Bell's inequality, 17, 107
Berry, Thomas, 129, 139
Beta waves, 116
Bierman, Dick, 51, 135
big bang, 99, 102, 103
block universe, 45, 46, 77, 80, 106
Bohm, David, 48–50, 73, 17, 22, 28, 48, 73, 100, 104, 106
Brownian motion, 63, 73, 77

C Series, 87
Cantor set, 75
challenge, 123, 127
clock time, 88, 94
coastline paradox, 58
coherent, 71, 72, 74, 78, 118
collapse, 17–22, 14–15, 7, 11, 23, 51, 68, 73, 84, 93, 97, 104, 105, 108, 111, 112, 114, 115, 116, 123, 129, 131
complex number, 76, 79
compression, 89–96, 83, 86, 100, 101
 lossless data compression, 92
computability, 10–11, 7, 15, 20, 58, 84
concrescence, 73, 104, 105, 106
condensation, 86, 88, 95
condensation velocity, 81, 82, 83, 84, 85, 86, 89, 90, 91
consciousness, 17–27, 6–10, 14, 15, 29, 32, 43, 50–54, 57, 58, 86, 87, 88, 96, 99–112, 115, 116, 123, 126, 130, 135, 136, 137, 138
continuous variables, 49
de Broglie, 13, 63, 69, 70, 85
de Broglie time, 69
de Broglie wavelength, 13, 69
delocalized, 71, 78
delta waves, 19, 114, 116, 117, 118
derivative, 64, 65
desired time, 124
dimension of scale, 3, 79, 91, 92, 93, 100
divergence, 62, 63, 67, 69
dopamine, 3, 90, 112, 118, 119, 120, 121,

122, 123, 125, 126, 128, 129
dreaming, 19, 22
edge effect, 67, 102
EEG, electroencephalogram, 19, 22, 114, 116, 118, 137
Einstein, 13, 14, 16, 37, 46, 135, 136
Einstein Podolsky Rosen (EPR), 11, 13, 16, 17, 34, 50, 52, 72, 107, 135
Einstein-Maric, 13
electron, 8, 9, 21, 34, 116
electron mobility, 9
electron mobility, 9, 21, 116
Elitzur-Vaidman bomb-testing problem, 39, 40, 42
entrainment, 115, 116, 117, 118, 122
entropy, 3, 25, 26, 27, 28, 31, 50, 96, 104, 106
Entropy, 25, 27
eternity, 50, 55, 111, 131, 133, 134
expanded moment, 51, 52, 111, 121
experienced time, 88, 119, 124
explicate order, 17, 48, 73
Feynman, 32, 33, 77, 78
fractal, 2, 3, 49, 61–108, 114, 115, 124, 130, 131, 134, 138
fractal dimension, 58, 60, 70, 74, 75, 81, 82, 83, 89, 90, 92, 96
fractal dust, 74, 75
fractal spacetime, 53
fractal prime, 83–86, 94, 99, 103, 108
fractal spacetime, 3, 53, 62, 63, 64, 65, 66, 68, 71, 85
frequency, 3, 13, 19, 21, 22, 26, 48, 89, 98, 111, 112, 113, 114, 116, 117, 118, 119, 123, 126, 128
gamma waves, 19, 114, 116
Gödel, 9, 37, 39, 84, 136
grasping, 84, 131, 132, 133, 134
Gravitational lensing, 34
Hameroff, 3, 7, 8, 10, 15, 18, 19, 20, 21, 22, 23, 50, 54, 105, 108, 115, 126, 137
Hameroff, Stuart, 8, 9, 19, 20, 21, 22, 23, 24, 105, 108
Heisenberg's Uncertainty Principle, 11, 15, 74
Hertz (Hz), 113, 118
hidden variables, 16, 17, 42, 107
Hoagland, Hudson, 115, 137
hydrophobic, 8, 9, 21
imaginary numbers, 75, 76
imaginary time, 46, 53, 77, 78, 79, 80, 137
implicate order, 17, 48, 50, 59, 73, 84, 135
insight, 3, 7, 9, 24, 53, 57, 80, 83, 84, 86, 94, 96, 131
integral, 39, 54, 73, 74, 77, 101, 109, 134, 137
intention, 94, 95, 96, 109, 110, 112, 117, 118, 123, 124, 125, 127, 130, 133, 134
interference, 11, 13, 40, 97, 129
Jaroszkiewicz, 76, 77, 78, 80, 137
Kant, 29
Kant, Immanuel, 87
Koch snowflake, 54, 55, 60, 61, 82, 83, 89
levels of description (LOD), 81, 82, 83, 91, 100, 112, 118, 124, 127, 129
light cones, 36, 37
lightlike separations, 36, 38, 43
logarithmic, 91
London dispersion forces, 8, 15
Mandelbrot set, 10, 49, 54, 56, 57, 58, 59, 60, 67, 68, 70, 76, 79, 89, 134, 135
Max Born, 14
McTaggart, 3, 87, 138
McTaggart, John, 87
Meck, Warren, 119, 122
meditation, 22, 130, 131, 134, 137
memory, 19, 32, 45, 82, 84, 85, 86, 89–97, 100, 101, 105, 109, 110, 111, 112, 115, 124
menstruation, 126
mental acceleration, 122
Merlin, 31
Microtubules, 8
Minkowski, 45, 46
Mobius strip, 111
momentum, 13, 15, 16, 18, 49, 63, 69, 113, 118, 127, 130
motivation, 119, 122, 125, 128

non-computability, 10, 11, 59
non-computable, 7, 10, 14, 17, 57, 58
non-differentiability, 61, 63, 64, 65, 66
non-local, 11, 17, 52, 72, 85
Nottale, Laurent, 3, 53, 54, 66, 67, 70, 71, 73, 78, 79, 62–81, 84, 85, 86, 92, 106, 138
null lines, 36, 38, 43
objective reduction, 17, 51, 112
Objective time, 87, 133
omnicentric, 103
omnigenetic, 103, 133
optimization, 66, 67, 92, 104, 106, 124
Orch OR, 18, 19, 20, 21, 50
Parkinson's, 120, 121
path integral, 77, 78
Penrose, 3, 7, 9, 10, 13, 14, 15, 17, 18, 19, 20, 21, 24, 25, 26, 34, 39, 42, 43, 50, 52, 53, 54, 58, 62, 80, 83, 84, 85, 94, 108, 112, 114, 115, 126, 131, 136, 137, 138
photoelectric effect, 11, 13
photon, 13, 32, 34, 36, 39, 40, 41, 43, 44, 49, 67, 72, 74, 84, 102, 111, 120
Plato, 7, 9, 43, 83, 101, 138
presponse, 51
Prigogine, 27
prolactin, 118, 125, 126, 136
psychedelics, 9, 22, 116
pyrogenic, 116
Pythagorean theorem, 45, 46, 59, 70
Quantum computing, 10, 15
quantum entanglement, 11, 50
quantum potential, 17
relativity, 3, 5, 7, 10, 18, 24, 25, 26, 30, 34, 44, 46, 49, 52, 53, 62, 63, 64, 70, 71, 75, 76, 107, 138
repetition, 57, 58, 86, 94, 95, 96, 97, 99, 100, 103, 111, 123, 129, 130
retrospective time, 88
reverse causality, 43, 47
Richardson, Lewis Fry, 58
Ritalin, 120
Sacks, Oliver, 90, 120, 126, 138
savor, 2

scale divergence, 62, 63, 69
Scale invariance, 92
scaling factor, 60, 61, 81, 82, 83, 89, 90, 91, 92, 95, 96
schizophrenia, 121
Schrödinger equation, 11, 13, 14, 77, 84
Schrödinger's cat, 14
self-awareness, 109, 111
self-similarity, 53, 55, 99
spacelike separation, 36, 37, 38, 43
spacetime blister, 18
speed of time, 3, 25, 26, 44, 88, 91, 97, 119
Subjective time, 87
superposition, 9, 11, 14, 15, 17, 18, 19, 20, 21, 24, 39, 40, 41, 68, 73, 97, 104, 105, 107, 114, 116, 126, 128
Superposition, 11, 14
surprise, 122, 123, 130
Swimme, 3, 129, 139
Swimme, Brian, 129
symmetrical, 2, 25, 28, 30, 31, 48, 73, 77
synchronize, 87, 100, 119, 121, 125
temporal density, 81, 82, 83, 88, 89, 90, 96, 118, 127, 129
temporal depth, 81, 82, 83, 89, 93, 100, 108, 123, 129
temporal flow, 3, 10, 25, 29, 44, 45, 47, 80, 83, 89, 91, 100, 102
temporal length, 81, 82, 83, 87, 89, 93, 108
temporal reversibility, 3, 20, 25, 32, 48, 52, 53, 62, 63, 68, 71, 73, 74, 80, 87, 105, 107
temporal symmetry, 25, 26, 29, 31, 32, 47, 48
thermodynamics, 24, 25, 26, 27, 28, 30, 50
Theta waves, 116
Thomas Young, 11
thought, 13, 16, 22, 23, 24, 29, 68, 85, 97, 105, 111, 127, 131
time dilation, 44
timelessness, 3, 7, 9, 10, 17, 18, 24, 25, 29, 30, 32, 33, 34, 43, 45, 47, 48, 49, 51, 52, 53, 57, 58, 67, 68, 71, 72, 73, 74, 79–111, 117, 118, 120, 126, 127, 128, 131
timelike separations, 36, 37, 38, 39, 42, 43,

52, 68
topological dimension, 58, 60, 74, 75
tubulin, 7, 8, 9, 15, 18, 19, 20, 21
Turing, 9
twin paradox, 121
ultraviolet catastrophe, 13
uncertainty principle, 13, 16, 18, 23, 32, 49, 69, 113, 125
van der Waals forces, 8
Vrobel, Susie, 3, 53, 54, 62, 83, 84, 85, 81–97, 99, 100, 103, 108, 124, 139
Watts, 23, 128, 139

wavelength, 15, 40, 98, 99, 113, 114
wave-particle duality, 11, 13
Weierstrass function, 65, 66
Wheeler, John, 34
whirlpools, 27
Whitehead, Alfred North, 19, 73, 104, 105, 106, 137, 139
Wick rotation, 76, 77, 78, 79
Wiener integral, 77, 78
Wiener process, 73, 77, 78
Woodward, Donald, 122

www.ingramcontent.com/pod-product-compliance
Lightning Source LLC
Chambersburg PA
CBHW021111080526
44587CB00010B/477